MAPS OF MY LIFE

MAPS OF MY LIFE

Guy Browning

◨ SQUARE PEG

Published by Square Peg 2008

2 4 6 8 10 9 7 5 3 1

Copyright © Guy Browning 2008

Guy Browning has asserted his right under the Copyright,
Designs and Patents Act 1988 to be identified as the author of this work

This book is substantially a work of non-fiction based on the life, experiences and recollections
of the author. In some cases names of people, places, dates, sequences or the detail of events
have been changed to protect the privacy of others. The author has stated to the publishers
that, except in such minor respects not affecting the substantial accuracy of the work, the
contents of this book are true.

First published in Great Britain in 2008 by
Square Peg
Random House, 20 Vauxhall Bridge Road,
London SW1V 2SA

www.rbooks.co.uk
Addresses for companies within The Random House Group Limited can be found at:
www.randomhouse.co.uk/offices.htm

The Random House Group Limited Reg. No. 954009
A CIP catalogue record for this book
is available from the British Library

ISBN 9780224082723

The Random House Group Limited supports The Forest Stewardship
Council (FSC), the leading international forest certification organisation. All our titles that
are printed on Greenpeace approved FSC certified paper carry the FSC logo. Our paper
procurement policy can be found at www.rbooks.co.uk/environment

Typeset in Adobe Caslon by Palimpsest Book Production Limited,
Grangemouth, Stirlingshire

Printed and bound by Firmengruppe APPL, aprinta druck,
Wemding, Germany

For the Fatted Calf,
who turned out all right in the end

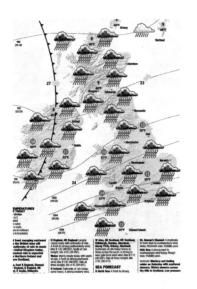

Acknowledgements

I would like to thank my mother and father, without whom this book (and my life) would have been a lot less interesting. I would also like to thank my wife, Esther, for the unwavering and loving support which I am assured will soon be forthcoming, and my children for sitting totally still while this book was being written.

At Square Peg and Random House, my thanks go to Rosemary Davidson and Mary Instone for their gentle editorial direction and for being so good as to publish this book in the first place. My thanks also to Matt Broughton, 'Bish' and Lily Richards for their help with drawing, designing and clearing my bizarre selection of maps. The hugely talented and shockingly modest Jenny Woodford helped me prepare some of my very early home-made maps, so a big thank you to her.

I'd very much like to thank the team at ICM Books – Laura Sampson, Lawrence Tejada, Margaret Halton, Karolina Sutton, Daisy Meyrick, Liz Iveson and Natalie Mason – who make everything about books apart from their writing an absolute pleasure.

Most of all, my thanks go to my agent at ICM Books, Kate Jones, who tragically died while this book was being written. Aside from being a truly magnificent human being, she was the compass who helped me steer my own literary course, despite it being well off the beaten track. In her own way she was a map-maker and trailblazer.

Maps and Me

I love maps. I can sit and read them for hours. My favourite reading in the small room is the *Philip's School Atlas*. There's something truly edifying about finishing your business in the bathroom and coming out also knowing where Burkina Faso is, or Tierra del Fuego, or Yorkshire.

I don't plan exotic trips when I'm spending quality time with my *Philip's School Atlas*; I just like to know where everything is so that when I meet an Ecuadorian I don't start talking about the plight of Africa in general. Equally, I've never been to New Zealand but if I'm ever lucky enough to get the chance to go, I like to think that I would head off in the right direction.

Maps are things of great beauty. They are all testament to our eternal quest to know what is beyond the horizon, to know how to get there and how to get back when beyond the horizon turns out to be not very nice. I like maps of the world, especially the ones where Britain seems to be bigger than Greenland. I also like large-scale maps of where I live that show footpaths and hedges and back gates to pubs. They're all good.

This book is about places and maps that made an impression on my early life. The history that goes with it might be a little dodgy but the geography is spot on.

Guy Browning
Oxfordshire, England

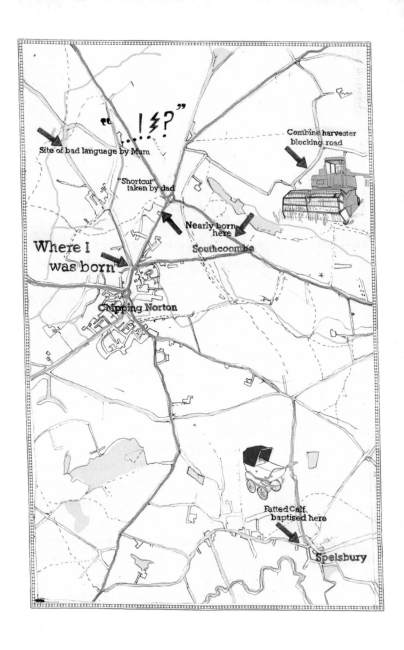

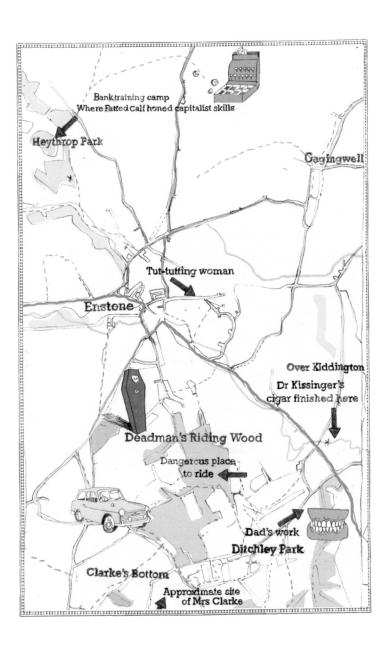

CHAPTER I

Chipping Norton, Oxfordshire, England

I was born into a two-parent family, which is like a one-parent family only twice as bad. I wasn't the first child. It's difficult to write that and somehow not feel that you've failed in something important. Second, as they say in sport, is nowhere. The firstborn is the golden fatted calf except it never gets slaughtered (except in really dysfunctional families). Second children are like second marriages: you know that in general they're worth having but you've also learned from your first how messy and unpleasant they can be. You'll notice that the announcement 'I'm pregnant again' never seems to get the attention of 'I'm pregnant'. Fathers tend not to announce to busy station concourses at the top of their voice 'I'm a dad for the second time!'

My older brother, the Fatted Calf, arrived thirteen months before me. Knowing what I know about him now, I am sure he was involved in some pre-birth negotiation with the powers that be that he should be born first and if that were arranged to his satisfaction there would be some hefty advantage for them. If these negotiations were anything like the negotiations he had with me for the next twenty years of my life, they would have resulted in him being given first slot as requested and the powers that be going away confused, upset and empty-handed. In a past life my brother must have been some kind of merchant adventurer. He was born to trade, haggle, increase margin, add value and generally redistribute wealth in his favour. My parents tell me that before I could talk he attempted to trade me for a set of felt-tip pens, albeit one with an extraordinary forty-two different colours.

I don't know much about those thirteen months when the Fatted

Calf was in the world without me, mostly due to the fact that I was developing from a two-cell embryo into something more closely resembling a human being. My parents, however, had a very clear idea of what to do with him during those difficult early months. Every morning when the Fatted Calf woke around six o'clock, they used to put a big lump of cheese in his hand and then pop him outside in the old black pram, with a protective squirrel net over the top. Then they used to go back to bed for an hour and get some more sleep, or attempt to conceive me, depending on how awake they were. I've looked through many childcare books since then and I still haven't found a reference to the cheese and squirrel net approach. Maybe it's a little bit of old Oxfordshire folk wisdom.

On the positive side, although I wasn't born first, I was nearly born in a stately home. Ditchley Park is a lovely old Cotswold-stone house on the road between Oxford and Chipping Norton in northern Oxfordshire. The house was built by Joshua Brown in 1723 who'd made his money by developing an early set of dentures in that heady time between the introduction of cheap sugar and the arrival of toothpaste. In the eighteenth century your money had to come from landownership for you to be socially acceptable and any form of trade was seen as brutish and unpleasant (my brother wouldn't have gone down too well). Being big in false teeth was therefore totally unacceptable even though it was only the rich who could afford them. Joshua Brown wanted to be socially acceptable and spent a lot of time gnashing his own teeth over this tricky problem.

His solution, eventually, was to choose as his wife the daughter of a poor but exceptionally blue-blooded aristocrat, the Duke of Accleugh. Her name was Becky and it turned out she had a better business brain than old Joshua himself. She urged her new husband to get into glasses, which he did with even greater success than teeth. Ditchley Park's association with bodily health (and the disintegration thereof) continued because in the early twentieth century it was bought by the Wills family, who'd made their fortune from tobacco with brands like Woodbines and Player's. If you go to Ditchley Park now you'll see a little smokers' shelter outside the main hall. What poor old Mr Wills would have made of this, I don't know. He's probably rolling himself up in his grave.

By the time my parents arrived at Ditchley Park it had become a conference centre specialising in very high-level and top-secret meetings between international leaders. On the estate, we lived slightly downhill from the main house in a small cottage imaginatively named the Lower House. My father's job at Ditchley Park, as far as I can make out, was to pick up very high-level people from Oxford station and convey them in conditions of extreme secrecy to Ditchley Park. I know this because my grandmother told me he often used to call in on her on the way back from the station if my mother needed something picking up or dropping off. In her time my grandmother met most of the world's leaders and power brokers and would have had an amazing collection of photographs with them had she not generally been taken by surprise in her curlers.

One of the very important politicians my father had to pick up from the station was the then American Secretary of State, Dr Henry Kissinger. He won the Vietnam War, or lost it – I can't remember which – but I'm sure if I gave him a lift from the station today he would quickly put me right. Putting people right was another one of his great talents. He was one of those people who was never wrong and even when they were clearly and obviously wrong they somehow made wrongness out to be the new right. Anyway, he was a very important politician though possibly not as important as he thought he was, but then show me a politician who is. Dr Kissinger was really important but he thought he was colossally important. If only he'd settled for really important he might have been a nice man, but he didn't and he wasn't. But that's not important for this story, and nor is Dr Kissinger which cheers me up because I know that Dr Kissinger would feel a slight pain at the knowledge that there were stories being told in which he wasn't very important.

Now I think about it I realise that it's only Dr Kissinger's cigar that's important in this story, although the cigar probably wouldn't have been important had it not been drawn through the lungs of the great man, the same lungs that released air conveying some of the most important American foreign policy decisions of the second half of the twentieth century. On the half-hour drive from the station, Dr Kissinger smoked one of his large cigars. It would be interesting to know whether it was a Cuban cigar because Dr Kissinger was responsible for creating the Cuban Missile Crisis, or solving it, one of the two, and at that stage they thought they could tackle the Cuban menace by forbidding the import of Cuban cigars. The modern equivalent would be trying to tackle al-Qaeda by forbidding the import of dates.

Dr Kissinger probably asked my father whether it would be OK to light up in his little car, an old Austin A40. For Dr Kissinger my father's car would have seemed like a quaint and slightly under-powered golf cart. For that matter, the whole of England probably seemed to Dr Kissinger like a quaint and well-tended golf course. My father would probably have said yes immediately, hoping that Dr Kissinger had a couple of fat ones in his suit and would hand one over. We know this probably didn't happen because the story only makes mention of the butt of one cigar.

Here I should make clear that my father doesn't smoke and hasn't smoked since he married my mother, who is convinced that cigarettes are the work of the devil and smoking them is akin to satanic worship. After marrying my mother my father developed, almost overnight, an enormously healthy addiction to fresh air. At intervals of an hour or even less, my father had to have his fix of fresh air. It didn't matter whether it was raining, snowing or nuclear fallout outside, nothing would get between my father and his fresh air. He was also a connoisseur of fresh air so that the whole family could actually be in the fresh air but he would know where to get some even fresher air somewhere else, generally out of the direct line of sight of my mother. My father's dedication to fresh air was legendary. He would find any excuse at any time, anywhere, during any function, however critical, to go in search of fresh air. I can only assume that his continual consumption of fresh air over the years must make him one of the healthiest men alive with lungs as pink as a cat's tongue.

Both my parents were very keen we didn't grow up to be smokers. One day – I must have been about twelve or thirteen – my father came in from somewhere (reeking of fresh air) and announced that smoking was a filthy habit and that he would give us children £100 each if we weren't smoking by the time we were eighteen. The Fatted Calf, who was already on twenty a day, decided that it would be fine to quit the week before his eighteenth birthday, suck on a couple of mints and claim the booty. He also tried to claim the £100 on the spot so that he could invest it until he was eighteen, calculating that the compound interest would result in him being about £2000 up on the original deal on offer. Sadly for him, the original deal was the only one on offer. My father forgot all about this promise until my eighteenth birthday when I reminded him of it with, quite possibly, a smug look on my face. After a moment of reflection and some fresh air he paid up, which I have to say was pretty impressive, especially as you could get a hell of lot of cigarettes for £100 at the time.

Of course, when my father was growing up, cigarettes were good for you, made you look good in front of women and were generally a friend for life. The Government itself regularly warned that not smoking would probably kill you and would certainly lose us the war.

Naturally kids took up smoking as soon as their pocket money would allow it. My father had his first cigarette up a tree when he was aged eight. He also had his second, third, fourth and twentieth cigarette immediately afterwards until he fell out of the tree, half-unconscious. This didn't put him off smoking although he's now very wary of trees.

Meanwhile, back at the Lower House in Ditchley Park, my mother went into labour shortly after breakfast. Knowing my mother, I can guarantee the labour wouldn't have been allowed to start until the breakfast things had been cleared away. The Fatted Calf had taken a very long time coming (and still takes an incredibly long time to have a shower – maybe there's a link there?) so no one panicked. My father went off to pick up Dr Kissinger from the station, called in at my nan's to pick up a tapioca pudding, a particular favourite of his, and drove back in a thick cloud of cigar smoke. My own labour was a very quick business (I take less time in the shower than it takes to flush a toilet, so there clearly is a connection). By the time Dr Kissinger had lit up his whopping Cuban cigar and was kippering my father on the A44 back to Ditchley, my mother was beginning to think that the car would be needed to get me (and her) to the hospital rather soon.

When Dr Kissinger finished his huge cigar he dropped the butt on the floor of my father's car. There was probably an ashtray in my father's little car but it's highly unlikely it would have been large enough to cope with the butt of Dr Kissinger's Montecristo, which would have been as thick as a standard cucumber. As Dr Kissinger unpacked his large and important personage from the car my father was probably thinking two things: one day that butt might be worth a fortune as a memento of the great man; and, possibly more importantly, it could be relit for at least a couple of good drags. This thought was interrupted by the arrival of my mother who by this stage was considerably larger than Dr Kissinger and, incredibly, more demanding. Barging the great man out of the way, she got in the car, threw the cigar butt on to the gravel, and we left for the hospital leaving the great Dr Kissinger to make his way totally unescorted to the front door of the house, probably a first for him. My father wasn't in a position to complain but subconsciously I think he's always regretted those two missed puffs.

Fortunately, my father didn't have to wait long for his next breath of fresh air. What followed was an incredibly fast and hair-raising drive to the hospital allowing for the fact that the top speed of his car was 54 mph and limited by the fact that it would disintegrate at speeds above that. We were also held up by a combine harvester on the A44 although I'm not sure how much faster we would have been travelling than the combine harvester in question. My father, despite knowing the way to the hospital fairly well, decided that this might be the right time to try another of his famous short cuts. In response to this, my mother used one of only three expletives she has ever used and brought a rapid stop to that nonsense.

Within half an hour, we arrived at the little cottage hospital in Chipping Norton and my parents rushed in. Minutes later my father came roaring back out of the hospital in search of some fresh air. A painter working on the outside of the building, who had seen them rush in, told my father not to worry and that everything would turn out all right in the end. My father reassured him that everything had already turned out all right and that Guy Montecristo Browning was now firmly established in the world. If you look carefully at the front of the hospital by the black guttering on the top left-hand corner, you can still see the accidental splash of white caused by the momentary shock of the painter. So on 19 August 1964 I was born in Chipping Norton and, as I was to discover some time later, that's not a bad place to have on your passport.

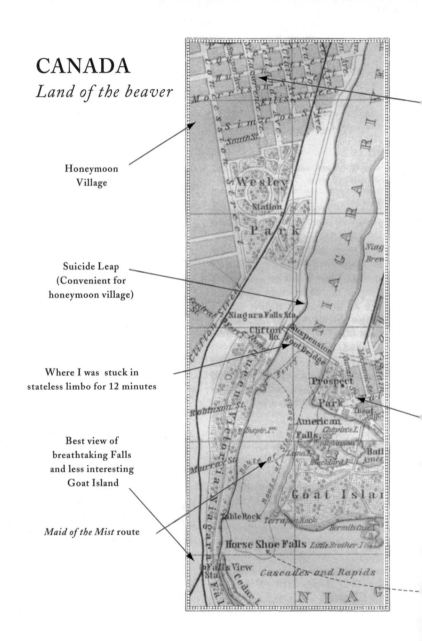

CANADA
Land of the beaver

Honeymoon
Village

Suicide Leap
(Convenient for
honeymoon village)

Where I was stuck in
stateless limbo for 12 minutes

Best view of
breathtaking Falls
and less interesting
Goat Island

Maid of the Mist route

NIAGARA FALLS AND

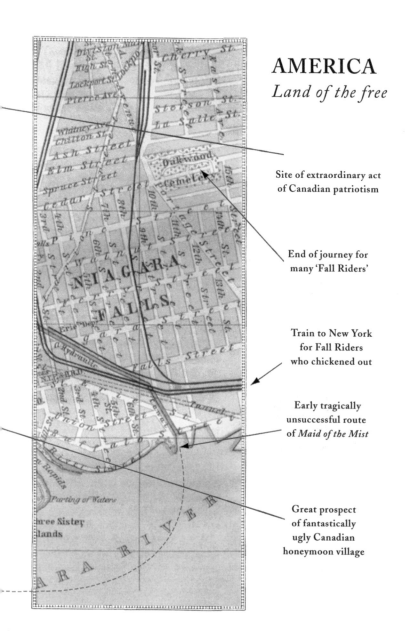

AMERICA
Land of the free

Site of extraordinary act
of Canadian patriotism

End of journey for
many 'Fall Riders'

Train to New York
for Fall Riders
who chickened out

Early tragically
unsuccessful route
of *Maid of the Mist*

Great prospect
of fantastically
ugly Canadian
honeymoon village

NEIGHBOURING COUNTRIES

CHAPTER 2

Niagara Falls, Canada

Nineteen years after being born in Chipping Norton I found myself at Niagara Falls, which is much like Chipping Norton except for the fact that it has a huge river crashing with phenomenal power over an awesome curving cliff in the middle. Chipping Norton also has a small stream running through it. I won't pretend that it tumbles over a cliff at any stage, but it is lively enough to power a thriving mill. The stream then goes on to join the River Evenlode a little way downstream which, as its name suggests, is probably one of the smoothest flowing and least demanding rivers in the county. The River Niagara also forms the border between Canada and the United States, whereas Chipping Norton is about five miles from the heavily fortified Oxfordshire/Warwickshire border.

I was on the Canadian side of the Falls which consists largely of a honeymoon village. Niagara is a very northerly and chilly place for Americans but a balmy, tropical paradise for Canadians. That's why they have a honeymoon village there because for them it is like Monte Carlo or Brighton. The village itself was seriously tacky and depressing, as if it had been built to make the rest of married life seem cheery in comparison. Every hotel advertised its prices on a board outside and I could never work out why these included single rooms unless they were trying to attract the trial-separation trade as well as the honeymooners.

I remember being very impressed that almost all the hotels advertised waterbeds. I've never understood why waterbeds should be an exciting thing to have. To my mind, at the end of a hard day you want to go to bed not to sea. When I turn over I want to get slightly more comfortable, not feel like I've just launched myself from a jetty. As for making love on a waterbed, surely that's not very sensible. When you're doing exact positioning work you don't want anything to be moving unless you or your partner have specifically requested it to move. Also, once you've spent the day staring at the Falls, you'd inevitably develop the subconscious fear that at some stage during the night the contents of your waterbed would plunge over the end of your bed, washing you and your beloved down to reception. Fortunately I wasn't on honeymoon and I didn't stay anywhere with a waterbed.

My abiding memory of Niagara was water, not surprising you would have thought given the giant rivers, waterfalls, waterbeds and so forth. You're probably thinking my abiding memory of the desert would be heat and dust and that I'm never going to make it as a travel writer. But the reason I remember water, apart from the Falls and the waterbeds and the colossal glasses of iced water served up in restaurants as a side dish to everything you ordered even if it were just a glass of water itself, was the fact that it rained as if the whole of Niagara and the surrounding countryside were themselves at the foot of a giant waterfall.

I was in Niagara for three days and it poured for three of them; really heavy-duty, serious, wet, thick, po-faced, industrial-strength rain. I discovered that Niagara Falls is much less impressive in the rain, especially heavy rain: it's just another downpour. You don't really want to ponder the magnificence of 40,000 tonnes of water per second going over the edge of the Falls when an equally large amount seems to be going down the back of your neck. It was the kind of weather that made you want to go indoors. Had there been a Niagara Experience multimedia show with plush seats in a dark room I would have been in there like a shot. But in those days air-conditioned multimedia shows were few and far between and you were forced to experience the actual things themselves outside, whatever the weather.

Apart from the seriously wet falls, there's precious little to do in Niagara unless you're on honeymoon and if you are you've probably got better things to look at. If the river hadn't chosen suddenly to drop one hundred and fifty feet at Niagara, you probably wouldn't think much of the place. The main tourist thing to do was to take a little boat called *Maid of the Mist* that sailed upriver as close as it could to the Falls themselves, generally disappearing into a cloud of spume and spray. All the tourists normally have great fun donning the bright yellow oilskins to keep them dry in the mist and spray. The day I was there the *Maid of the Mist* was doing very slow trade as you would have had exactly the same experience hopping on a local open top bus which would then instantly have become *Bus of the Mist*. The few tourists who did decide on the boat trip were already wearing waterproofs when they boarded or were already soaked to the skin. You could see when they traipsed off the boat an hour later that all they were thinking of was their hotel rooms, getting their wet things off and slipping into their lovely dry waterbeds.

Instead of wasting my very limited budget on the *Maid of the Mist* I decided to go to the cinema to dry out. It was a Monday afternoon and it was raining. There was no one in the small cinema and I sat at the back by myself ready to be culturally uplifted by *Rocky II*. I don't remember anything about the film except it probably had two men hitting each other very hard but I do remember exactly what happened before the film started. Just before the curtains went back or up or, probably, given the fact that it was Niagara, down, another man came into the cinema. He didn't see me at the back and plonked himself down in the middle of the front row.

The curtains went back and a giant picture of the Canadian flag appeared on the screen. It's a lovely flag the Canadian flag – big, bold colours but with a nice leaf in the middle.[*] Seconds after the flag appeared on-screen a tune started up which I took to be the Canadian national anthem. It's not instantly recognisable in the same way as the US, German or French national anthems. If it's not too late I would suggest to the Canadian authorities that they change it to 'Land of the Silver Birch, Home of the Beaver' as this is a great tune with a cracking chorus: boom-ti-di-li-aye-day, boom-ti-di-li-aye-day, boom-ti-di-li-aye-day, boom. You can also do canoe-paddling movements with your arms which is an added bonus. Americans put their hands on their hearts during their national anthem, which I think is overdoing it slightly.[†] No sooner had Canada the Brave or Beautiful or Leafy started than the gentleman in the front row stood up and remained standing for the duration of the anthem. It could be that he'd dropped some popcorn but I think he was actually showing respect for the flag.

I was still at the back quietly nursing my Twix and was extremely impressed by this unforced show of patriotism. On reflection, it was a perfect example of the old adage (adage sounds like a kind of dull, green vegetable: 'Not adage again, Grandma!'), a perfect example of the old adage that a gentleman is a man who uses a butter knife when alone. The Canadian gentleman in front of me thought he was

---

[*] See Chapter 13 for my favourite flags. Canada did amazing work in the Second World War, especially on the beaches of Normandy – see Chapter 17 for my personal invasion of Normandy – and our Royal Family would now be resident in Canada if we'd lost the Battle of Britain. That would have given the French Canadians something to think about, being back under the direct rule of a British monarch after a gap of six hundred years.

[†] See Chapter 13 where I come close to being lynched for not applying hand to heart during the American national anthem.

all alone, certainly didn't know that he was being watched but still did the right thing. I realised about three bars into the national anthem, and one bar into my Twix, that perhaps I should also stand up, not just to respect Canada and its national anthem but also to show some respect for the man who was showing some respect.

I think I was still working out my correct response and whether butter knives should be used, when the anthem stopped, the man sat down and the adverts for local drive-in muffler and shock places started. When the film was over I left the cinema, looking closely at the other man to see if he displayed any outward signs of patriotism. I don't know what I was expecting – perhaps a big tattoo of a moose on his neck – but he was outwardly very normal, which I suppose is very Canadian in its own way.

It was still raining when I left the cinema. I wondered whether Niagara ever worried about flooding. It would be good to see a sign saying 'Niagara Falls Closed Due to Flooding'. It's not beyond the realms of possibility as British railways frequently come up with even better headlines such as 'Delays Due to Train on Track'. The only other place open in the town where you could walk round in the dry and not spend any money was the Niagara Falls Museum so I wandered off in that direction. The museum was a big old factory on two floors. Exhibits in museums of the time weren't interactive unless you happened to knock one over. There were no buttons to press, no video booths, no dioramas, no experiences and no guides dressed in authentic but ridiculous costumes. Instead the whole museum was stuffed to the rafters with a whole dusty mass of assorted clutter. Had my mother been there she would have taken one look inside, stormed off to the director and told him to go and tidy up his museum as it was an absolute disgrace.

The museum didn't need much help to be interesting because it had such a bizarre combination of exhibits. Downstairs they had a collection of the barrels in which various daredevils, madmen and suicidal maniacs had attempted to go over the Falls. I remember these vividly partly because in at least half of them the occupant had died, either by drowning, suffocation or being shaken to death like a giant cocktail. There was one device that looked like a heavy industrial boiler covered with big manly rivets. Inside there was red cushioning like an Edwardian railway carriage. Clearly a lot of thought had gone into this one which was a shame because it was caught under the Falls and the occupant was eventually shaken to death. There was also a big rubber ball which was a bit odd. Did the owner think he was going to bounce off the bottom? We'll never know because he didn't. And then there were the tin baths and water wings and giant carrier bags that the seriously cracked pots had used to go over. I think they had as good a chance of surviving as any of the men and women in the more heavily engineered contraptions.

Upstairs in the museum the collection got even more bizarre. The whole floor was given over to a collection of deformed animals: sheep with eight legs, cows with three heads, Siamese horses etc. It was almost as if the museum was trying to say, OK we humans do some pretty stupid things in barrels downstairs but look at these idiotic animals upstairs. Whatever they were trying to achieve, it was pretty sick-making and I left feeling disgusted and depressed. It was still raining as I trudged back to my hostel. As I walked past the honeymoon hotels I thought that the combination of waterbeds, that unpleasant museum and the driving rain was not going to be a good start for any of the young couples. But then I thought of Blackpool and suddenly it didn't seem quite so bad.

The following day the rain had cleared sufficiently for falling water to seem a slightly more exciting tourist attraction. I started off by walking down to the observation platform on the Canadian side where you can get very close to the edge of the top of the Falls. This is not as exciting as it sounds because if you look one way there's a big river and if you look the other way there isn't. After a while I decided that you'd actually get a better view of the Falls from the American side and I set out to cross the Rainbow Bridge, which arches high over the Niagara Gorge just downstream from the Falls themselves. I didn't have an American visa in my passport but my father had drummed into me that borders were only lines on the map and I thought that visa restrictions could be waived in my case because of this strongly held opinion which I now also shared.

I made my way along the road until I reached the Canadian side of the magnificent Rainbow Bridge. The Canadian end was guarded by a little customs post. Entering the post after politely knocking on the door, I found a single customs man sitting inside, with his feet on his desk and his eyes hidden behind the mirror lenses of his sunglasses. He may also have been chewing a toothpick, although I might be imagining this last bit. He certainly talked in a manner that would have retained a toothpick or small cigar or even a large sausage firmly in one side of his mouth.

Looking me straight in the eye (actually he might have had his eyes crossed for all I know behind those mirror lenses), he shot one word across the room. 'Birthplace?'

I knew in the millisecond before I answered that none of his advanced training at the Royal Mounted Customs Academy would equip him for what was to come. 'Chipping Norton,' I said in the cool manner of an RAF pilot releasing a bouncing bomb across a lake towards a dam in the heart of Germany. I would very much like to say at this point that the toothpick dropped from his mouth but as I can't be sure it was in his mouth in the first place, I can't really claim that it then fell out. Nevertheless, in that second you could have heard a toothpick drop.

He asked me to hand over my passport so that he could read for himself that this wacko place existed. Satisfied that my birthplace was something called Chipping Norton (which, come to think of it, to Canadian ears must sound like a leathery old woodchuck) he handed my passport back and I explained to him that I wanted to cross over to the American side of the Falls to see what they looked like from that side. He looked hard at the space around my feet (he could still have been cross-eyed behind those glasses) and noticed that I had no luggage, which seemed to reassure him. He said as long as the Americans on the other side of the bridge didn't mind letting me in, and as long as I was back in Canada before the end of his shift, that would be OK by him.

I thought this was very open-minded and reasonable of him and I began the short walk across the magnificent bridge, feeling not unlike someone crossing Checkpoint Charlie at the height of the Cold War (which was still very hot then). I began to have second thoughts halfway across the bridge when I looked back at the booth on the Canadian side and saw the Immigration person still staring (cross-eyed) at me but this time with a big smile. From the middle of the bridge I could make out the American customs post on the other side with a similar uniformed figure within. I suddenly realised that I was in stateless limbo. I could become like that poor man stuck in an airport lounge somewhere unable either to travel anywhere or properly arrive. Except I would be much worse off because he had washing facilities and restaurants and seats to sleep on and vending machines and interesting security announcements. I just had a stretch of asphalt and a good view of Niagara Falls if the weather ever cleared up. I could be trapped there for years, grow a long beard and become a tourist attraction in my own right.

Eventually I gathered myself together, mentally claimed the centre of the bridge for Britain and made my way to the booth on the American side. I was more than a little anxious because even French waiters get upset at the rudeness of US customs officials. I walked into the American booth and word must have spread across the international news wires of my birthplace because he didn't even ask about it. He listened politely to my limited travel plans, asked if I intended to bring down the government of the USA by force (memories of the War of 1812 between the British and the US were clearly still fresh in his mind) and, when I assured him that I wouldn't have time for this as I had to be back by the end of his Canadian opposite number's shift, he stamped my passport with a one-day visa and waved me out into the free American drizzle.

It was twenty years before I recrossed the border. At least that's the kind of story I had in my mind as I wandered around the American side of the Falls, calculating precisely how far I could walk and still leave time to get back before the Canadian shift ended, factoring in some contingency time for eventualities, like a cup of tea. One of the thoughts I must have had that day was what the visa requirements were for going over the Falls in a barrel. I suppose that's something you sort out after you've been fished out, if it's still relevant.

On the American side of the Niagara Falls they have a national park which is very nice and green and lovely but slightly spoilt by the view of the Canadian side and its suppurating honeymoon village. Eventually, with an eye on my watch, I decided the view might be better from the Canadian side and made my way back across the bridge. As I passed my cross-eyed friend in the Canadian booth he opened the door and said, 'Welcome back, Chipping Norton.' It obviously made a big impression on him.

NOTE: I once had an American girlfriend who was similarly impressed by my birthplace. She was also fascinated by my liking for corned beef. Somehow the two got confused in her mind and she used to refer to my chipping beef sandwiches. Sad to say, the relationship petered out but there was something about it that did last. On a trip to the Middle East she brought back a gift for me in the form of a small trinket box made of inlaid wood with a red cloth interior. When I first opened it I was nearly knocked out by a cloud of perfume so thick and over-whelming it felt like a genie of great power and confused sexuality was blasting off from inside. To remind me of her, she'd put a dab of her perfume into the box. I can only assume the top of the bottle came off at the vital dabbing moment because that little box was like a police wagon full of tear gas. Almost half a century later the memory of that girlfriend grabs me by the throat every time I select a cufflink. I'm sure the sight of corned beef has the same evocative effect on her.

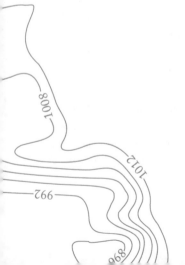

CHAPTER 3

SAN SALVADOR, EL SALVADOR, CENTRAL AMERICA
MY FIRST DIPLOMATIC POSTING

Volcano conquered by
Father in Ford Consul

Birthplace of
Anglo-Salvadorian
perpetual friendship

Chipping Norton

Poor peasant -
rich zone

Relief flight of
porridge oats

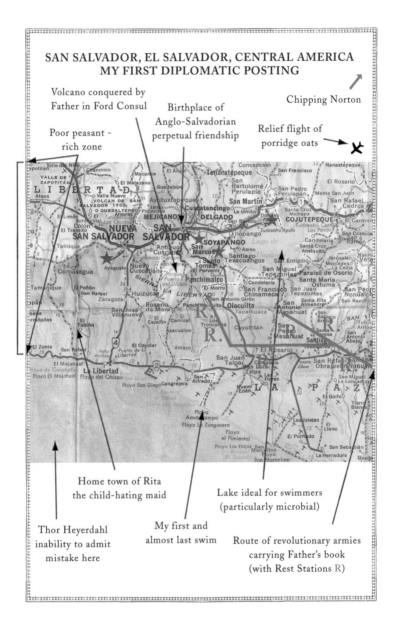

Home town of Rita
the child-hating maid

Lake ideal for swimmers
(particularly microbial)

Thor Heyerdahl
inability to admit
mistake here

My first and
almost last swim

Route of revolutionary armies
carrying Father's book
(with Rest Stations R)

San Salvador, El Salvador, Central America

Very few people know where El Salvador is. Even fewer people have been there. Many people who live there don't even know where they are. El Salvador is tucked away in the Pacific underbelly of Central America and although it's the smallest country in the region it's also the most densely populated, fortunately with charming, polite and generous-natured people. That's where I spent the first three years of my life and it is to that fine country that I owe my abiding love for all things Hispanic and my deep dislike of porridge.

The step from Chipping Norton to El Salvador is not an obvious one. Many people leaving Chipping Norton find Oxford or Banbury to be the logical next step but my father was never interested in logic or, for that matter, steps. Instead my father had a passion for taking a flying leap into inexplicable and seemingly random locations. If there was an odd place to be at an odd time then that place would be top of his itinerary. Places were important to him and not something around in which to be mindlessly swilled. He studied geography at university and was particularly interested in how places used their land and how this affected society. After his first degree he decided he wanted to do a doctorate in agriculture and economic development, which proves that he really was very interested in it (or that he was taking a long time to understand it).

It soon became apparent that countless other people had also studied this area. As academic fields go, it was bristling from hedge to hedge with researchers, experts and professors which was not altogether surprising because what they were all studying in effect was how people grow food to eat. Not one to give up easily, my father eventually found one small part of the world that hadn't been exhaustively studied and that was El Salvador, principally because no one knew where it was. He quickly worked out that there were fields there and a developing economy and almost no one studying them. All he needed was a job to pay for the family to be there while he did his studying. That's how he joined the Foreign Office and that's how I, the Fatted Calf and my parents found ourselves in San Salvador, home of the pineapple and capital city of El Salvador.

In his day job at the embassy my father got straight to work promoting El Salvadorian–British friendship and cooperation. This was largely a matter of explaining who the British were, where Britain was and why we suddenly wanted to be friends. Where, the Salvadorians quite rightly asked, were the British in 1528 when the Spaniards were laying waste to their country and they really needed us? My father told them that our King Henry VIII was having trouble with the wife/wives around that time and they seemed to accept this as a reasonable excuse. Having answered this and other questions to the best of his ability, my father would then slip away and work on his doctorate, looking closely at fields, who owned them and what was growing in them.

One of the things that tickled my father most about his new job in the embassy was that it came with an equivalent naval rank attached. It probably wasn't as elevated as commander or captain but it would definitely have been some kind of officer. This was a big step up for my father who had been in the navy during national service as a humble signalman. He wasn't very good at this for two reasons: the first was that he never really mastered Morse code, the second that he generally thought he could improve on whatever message the captain of the ship was trying to send. Of all the world's navies, the Dutch were generally considered to be the most efficient and their signalling the most rapid. When a Dutch warship was sighted, therefore, my father was generally hustled off the bridge before he could start making up signals and disgrace the Royal Navy.

Ten years after leaving the navy, his old ship called in at El Salvador on a tour of Central America. The embassy staff went to pay their respects and my father was ceremoniously piped aboard the ship as befitting his officer rank. While he was being shown around the bridge, a message came through in Morse giving details of the marital status of all the women in the embassy party for the benefit of the naval officers on board. My father translated this message aloud to the party as it came through. It was not the first time he'd been ushered off the bridge in disgrace but this time he'd managed it as an officer.

My mother, meanwhile, was busy at home recreating English domestic conditions in downtown San Salvador. We had a nice bungalow courtesy of the Foreign Office but the running of it and the family inside posed a few problems. What my mother needed more than anything else was a thick, no-nonsense magazine called *Good Tropical Housekeeping*. Much to her frustration there wasn't one, so she had to make things up as she went along. The first thing she tackled was the food. Her own mother had given dire warnings that a spell in the tropics would inevitably lead to rickets in the children because the local vegetables had no nutritional value and, even if they did, it would be of no practical use because the whole family would be bent double with diarrhoea for 90 per cent of the time.

In the street markets there were a few trustworthy British vegetables such as potatoes and tomatoes but a lot of other dubious local stuff that could take itself straight back to the jungle as far as my mother was concerned. My father gently pointed out that these 'British' vegetables originally came from Central America but my mother was of the unshakeable opinion that four hundred years growing in the healthier, richer soils of Britain had naturalised them and made them a far higher order of vegetable. The other weapon in her war against rickets, malnutrition, beriberi, bilharzia, diarrhoea and the long list of tropical ailments she kept pinned up on the fridge was porridge. Half the weight of the diplomatic bag which arrived at the embassy each week must have been porridge oats because that's what we had for breakfast every day for the three years we were there. Every morning the fresh pineapples that littered our back yard were cleared away and the table laid for porridge.

The other thing that my mother struggled with in the tropics was the wildlife, especially the large quantity of it that seemed to want to live inside the house with us. A whole rainbow of colourful lizards would sit on the ceiling with their lazy eyes watching my mother dusting or preparing vast cauldrons of porridge. Occasionally, through sheer laziness or stupidity, one of the lizards would drop off the ceiling often on the head of a small child. I know my mother found this trying and one of the first instructions she gave to our two maids was that a total exclusion zone for lizards should be established around our bungalow for a distance of approximately a mile.

Our maids didn't take this injunction too seriously possibly because there were a quarter of a million lizards in the proposed exclusion zone and also because I think they derived enormous entertainment watching lizards drop on our little blond heads; they may even have shouted encouragement. Latin Americans in general love their families and love little children (I know I'm generalising here but we English love generalising). My mother, on the other hand, didn't want us to be spoilt by continual displays of noisy affection, lest we inadvertently grow up to be Italians. Instead she hunted around for a more traditional style of English nanny, in other words one who showed no signs of liking or being remotely interested in children.

To her great credit she managed to find the only two women in the whole of Central and possibly South America who showed complete indifference to children and they became our nannies.

This recruitment strategy gave rise to my brother's first brush with death. One element of the local wildlife my mother particularly disliked was a little green spitting frog, possibly because it was and is one of the most ill-mannered and poisonous little buggers known to man. Where they'd got the spitting frog from I don't know but our two maids had somehow managed to position it directly in front of the Fatted Calf who must then have been about three. When my mother found them, the frog and my brother were gazing at each other calmly. The frog was probably deciding exactly where it was going to spit and my brother was deciding whether he could claim ownership of the frog and sell it at the local farmers' market. Meanwhile, the two maids were waiting to see whether the powerful porridge magic worked against the lethal frog poison. My mother rapidly put a stop to all this nonsense and inevitably there were repercussions. The morning after this incident the porridge ration was doubled.

Lounging around in crystal-clear, sun-dappled swimming pools plays a big part in any ex-pat lifestyle unless you're from another country and you find that you're an ex-pat in Britain; then the crystal-clear, sun-dappled swimming pools are replaced by rain-sodden jumble sales. In El Salvador the Browning family swam in three places. The first was the pool at the leisure club for employees of Shell and other people who were British or that way inclined. As this was the most natural and easiest place for us to hang out, my father tended to avoid it. Instead we often went down to the Pacific shore outside the city. Here the beaches were certainly beautiful and fringed with palm trees but they also happened to have the most powerful and dangerous rip tides on the entire Pacific coast. Many an epic ocean crossing was started off this coast by someone intending to go for a little trip across the bay. My suspicion is that the great Norwegian explorer Thor Heyerdahl only meant to have a few cocktails in view of the beach rather than try and sail to Polynesia but that the rip tides pulled him offshore in an instant and, like my father and Dr Kissinger, he couldn't admit that he'd made a mistake so he had to pretend he always intended to cross the Pacific on a bag of twigs. Similarly, a small child could be washed off the beach and out to sea before you could say, '¿Donde esta mi niño?' (Where is my child?) It wouldn't have surprised me if this beach hadn't been specifically recommended to my parents by our maids.

Eventually we found a safer place to paddle, a deep lake sitting in the crater of a long extinct volcano. The sand was a fine, jet-black sediment of lava, gently eroded by the small, wind-blown waves that rippled across the lake. There is a photo of us paddling on the shore of that lake and it does look strangely beautiful. Somewhere we also have medical records of the yellow fever we picked up from the water there which had my mother in bed for six months taking porridge intravenously. Shortly afterwards my father fell ill, too, with vicious bouts of diarrhoea. The embassy doctor asked him to send in a sample that he could analyse for anything nasty (nastier than the sample itself, that is). Looking round for a suitable receptacle, my father picked up the plastic container for a 35mm film. In due course he filled it with the required sample, popped the lid back on and it was sent off to be analysed. A couple of days later the confused doctor asked my father why he'd sent him a roll of film to be analysed. The only person more confused would have been the little man in the photo shop who tried to develop the contents of the other container.

One of the high points of our stay in El Salvador – quite literally – was my father's attempt to drive over a live volcano. My father, much like Dr Kissinger or Thor Heyerdahl, has never knowingly been in the wrong and this certainly applied to anything to do with navigation. El Salvador didn't really present much in the way of navigational challenges because there was normally only one road to choose from. This didn't prevent my father acting like any other man in similar circumstances and deciding that there must be a short cut to where we were going despite the fact that we were going from A to B and there was one road running directly between the two. In my experience roads represent the cumulative wisdom of generations of travellers who have found that there are no shorter cuts to where you want to go. My father clearly didn't share this view which is why we found ourselves driving up the side of an active volcano.

I should imagine there was some kind of goat track to begin with but that would have been all the encouragement my father would have needed. About an hour later, when we had probably driven above the level of mist that had initially obscured the huge mass of volcano in the direct line of our short cut, my father succumbed to the twin pressures of worried wife and melting tyres. Executing a tricky ten-point turn on a narrow lava ledge halfway up the volcano, my father then drove back down the volcano. It goes without saying that my father didn't drive back the way we had come up because he thought there might be a short cut back to the main road that would save us even more time.

That's how it happened that we emerged in our Ford Consul at the top of a track up which an entire village was processing in a religious ceremony designed to appease the gods of the volcano. Seeing our family emerge from the mist in our Ford Consul was in a small way tantamount to the arrival of ancient gods from a faraway planet and it must have been then that my father began to acquire his legendary status in the country as 'man who drives family up (and down) volcanoes'.

My father's genuinely legendary status in El Salvador really took hold after he'd finished his research, resigned from his job and we'd left the country crippling porridge oat exports for the year. On his return to Oxford my father wrote up his doctorate called *The Land and the People*, which sounds like a fantastically dull and worthy Ken Loach film but in Spanish is *La Tierra y el Hombre*, which sounds more like a spaghetti western and far more exciting. The basic gist of the book was that only a few people owned all the land in El Salvador and grew cash crops for export. The rest of the population didn't have land so basically starved. Everyone in El Salvador knew this already but no one had bothered to say it, except, now, the legendary Englishman who drove down volcanoes.

My father's book rapidly became the revolutionary handbook for the resistance movement that was beginning to emerge in the country. In a way it was like Mao's *Little Red Book* but much, much thicker and heavier. The slow pace of the revolution in El Salvador can largely be explained by the fact that my father's book was so heavy to carry around in the rebels' rucksacks. It was very noticeable that the pace of the revolution picked up markedly when the lighter paperback edition was published.

American foreign policy for the last fifty years seems to have been to assemble the world's most impressive set of weaponry and then to shoot itself in the foot. At this time American policy in Central America was to label any genuine popular movements by the poor and oppressed as Communist and then arm to the teeth the repressive leaders of the country to fight against Communism. This could well have been another one of Dr Kissinger's great ideas. My father, along with several million Latin Americans, found this policy rather foolish if not downright iniquitous. As a way of countering American influence in Latin America he would invite guerrilla leaders round for tea at our house. Having seen some of these guerrilla leaders in action around our tea table I got the distinct impression that they found the whole experience more confusing and intimidating than a long war in the hills. There is something about my mother insisting that another rock cake needs to be eaten that is not unlike hearing the approach of a helicopter gunship.

After three years, my father's research was complete and in his day job at the embassy he had secured eternal and undying friendship between the peoples of Great Britain and Northern Ireland and El Salvador.* We left El Salvador in more style than we arrived because it was still the right of any Foreign Office person to return home by sea. My father chose to fly up to New York and then across the Atlantic on the *Queen Elizabeth*. This was the first *Queen Elizabeth*, which ended her days in Hong Kong harbour in 1972 and is now incorporated into the foundations of the new airport, rather than the more famous *QE2*.†

There is a remarkable photo of me in the bows of this beautiful ship, in my blue anorak with my arms outstretched in exactly the same position as Kate Winslet on the *Titanic*. I suppose by doing that I proved that assuming that position in the bows doesn't actually cause a ship to sink. Even more instructive is the fact that the Fatted Calf is behind me, hunched over, counting what looks like change in his little hand. It was on this trip, so my parents tell me,

* Interestingly, my father always had difficulty explaining the Northern Ireland thing and the El Salvadorians had a lingering suspicion that this friendship they were entering into might be a bit of a *ménage à trois* or, as they say in Spanish, a *tortilla con tres huevos* (omelette with three eggs).

† Shortly after the Second World War the passenger liner *Queen Elizabeth* was crossing the Atlantic from New York when it met the British battleship HMS *Queen Elizabeth* coming the other way. The captain of the battleship sent a signal to the passenger ship: 'Snap!' Had my father been on the bridge it would have come out 'Pans!' which wouldn't have had quite the same impact.

that he discovered a) that coins were often left in the return slots of vending machines, and b) he could now reach those slots. I wouldn't say he collected enough over the five days to work his passage, but it probably gave him a capital base to fund his future ventures wherever we next found ourselves which, as it happened, turned out to be a village called Kidlington back in the golden navel of the universe, Oxfordshire.

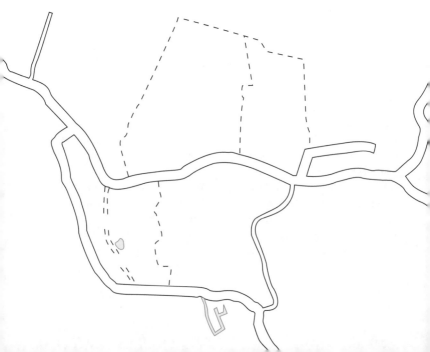

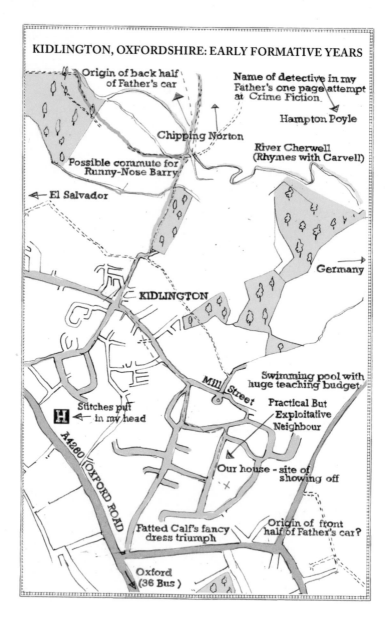

KIDLINGTON, OXFORDSHIRE: EARLY FORMATIVE YEARS

Origin of back half
of Father's car

Name of detective in my
Father's one page attempt
at Crime Fiction

Hampton Poyle

Chipping Norton

River Cherwell
(Rhymes with Carvell)

Possible commute for
Runny-Nose Barry

← El Salvador

Germany →

KIDLINGTON

Swimming pool with
huge teaching budget

Mill Street

Practical But
Exploitative
Neighbour

Stitches put
← in my head

A4260 OXFORD ROAD

Our house - site of
showing off

Fatted Calf's fancy
dress triumph

Origin of front
half of Father's car?

Oxford
(36 Bus)

CHAPTER 4

Kidlington, Oxfordshire

Runny-Nose Barry. When I think of Kidlington, he's the first thing I think about. I'm sure it's not what the village elders had in mind for the public image of Kidlington which can, after all, claim to be the largest village in England. There are no signs on the way in saying 'Welcome to Kidlington – Home of Runny-Nose Barry'. When I drive in, however, he's the first person I think of and I can't think of him without thinking first of his nose and second of his pedal car.

We moved to Kidlington when we got back from El Salvador. My father decided he didn't want to be in the Foreign Office. He was far too diplomatic ever to say why but he decided he would finish his doctorate instead and then become an academic at Oxford University, which was handy for my nan and tapioca pudding. My parents couldn't afford to buy a house in Oxford itself so they bought a little two-up two-down in a big estate in the biggest village in England – Kidlington. It's the biggest village because it has an enormous number of two-up two-downs but nothing of interest that would make it a town. It did, however, have a good bus connection to Oxford and, although my parents couldn't have known this when they put their deposit down, it did have Runny-Nose Barry.

Often when me and the Fatted Calf went to play in our little garden Barry would be parked at the end of our drive (four feet away), his pedal car blocking the entrance. He would just watch us, his hands never leaving the steering wheel and his nose gently running.

It was a classic stakeout position. I remember his pedal car in greater detail than I remember Runny-Nose Barry. It was blue, open-topped with some impressive fifties-style chrome work down the side, two doors obviously and a grill like an old Humber. There were some scratches down the side, but nothing that wouldn't polish out.

Neither me nor the Fatted Calf had a pedal car so the sudden arrival of Runny-Nose Barry was pretty exciting. Truth be known, we had no idea where he came from. We assumed he lived a couple of doors away but for all we knew he might have pedalled all the way from Oxford, leaving a long trail of mucus behind him. He was a mystery man, that Runny-Nose Barry. But, and I'm sorry to have to recall this, he was also dull. Runny-Nose Barry didn't have much going for him except his nose and his pedal car. I don't remember ever actually playing with him. I don't remember him ever getting out of his car. He would just sit there, hands on the steering wheel, nose quietly flowing. One moment he would be there, next he would be gone, maybe sitting at the end of some other kid's drive. Who knows? But he's the only thing I remember about Kidlington.

Well, not quite. I left Kidlington when I was six so it would be pretty disappointing if Runny-Nose Barry was the sum total of my memories. I also remember getting in trouble with Captain Scarlett, a super-hero very popular at the time with a super Corgi model car featuring a very pointy nose. Somehow, I had acquired one of these cars and I discovered that if you drove it very hard and fast into the skirting board in the living room, the pointy nose would leave a satisfying indentation. I think I must have been up to fifty or sixty satisfying indentations before my mother worked out where the curious low-level rhythmical thudding was coming from. I remember very little for a year or two after that which means either that I have a very poor memory or I had a colossal beating. I do know that thereafter all model cars entering the house were checked for crumple zones at the front end that would minimise the effect of head-on impact with skirting board.

The fancy-dress parade at my infant school was another high point of my time in Kidlington. All top executives have a black ink pen as thick as a farmer's finger for signing big contracts. The Fatted Calf secured his in Kidlington approximately twenty-five years before he became a top executive and needed to sign big contracts. Blenheim Road Infants School had a fancy-dress competition at the end of term and all of us children were encouraged to enter. As luck would have it we had a big rectangular cardboard box at home which suggests to me a large purchase must recently have been made. Purchases were few and far between in our house, especially those which came in a box, so I'm disappointed that I can't recall what came in it. Thinking about it, my younger sister turned up around that time so it could have been her. If it was, all I can say is that she came in very impressive packaging.

The Fatted Calf discovered that if he pulled the box down over his head, just his feet would be visible from the bottom. This, of course, was a blessed relief for the whole family and we all fervently hoped that he would be happy to stay like that for a few months. Instead he had the admittedly rather good idea that he could go to the fancy-dress competition as a skyscraper. My father flung himself out of the sofa, drew a couple of windows on the box and the painstaking preparation was complete. The Fatted Calf stood quietly in the box for a while, clearly pleased with the fact that his head was now in the penthouse suite area of the skyscraper where it felt naturally at home. I think he may have requested windows be cut to improve his view but this was vetoed on aesthetic and noise-abatement grounds.

I remember very clearly the Fatted Calf in his cardboard box moving forward across the playground towards the judges' table behind seventeen Captain Scarletts and fourteen Snow Whites. He must have been following the noise of the kids in front of him and watching the two inches of playground tarmac in front of his feet, because he still didn't have any windows. I can also guarantee that, when his cardboard box finally arrived in front of the judges' table, the Fatted Calf would have been smiling his winning salesman smile. The thing about that salesman's smile is that it works even through thick cardboard and, sure enough, my brother walked off with the first prize of that fat executive pen. It'll come as no surprise to you that I can't remember what I was dressed as. I probably went as a thick ear if the competition was at any time after the skirting board incident. Of course everyone remembers the skyscraper because that's what happens with skyscrapers: when you're inside one you have a great view but the only thing everyone else can see is a great big skyscraper.

Given the fact that my brother was beginning to pull off great PR stunts like the skyscraper, I think it must have been around this time that I started showing off with a vengeance. One of the pieces of furniture we had in our little house was a wooden coffee table. Our furniture in Kidlington came from two places: it was either very light, flimsy stuff brought back from El Salvador or it was very heavy, solid stuff made by German POWs. The end result was a domestic style akin to the German Embassy in Bogotá, which was certainly not a common look in the village or anywhere else at that time.

Let me quickly tell you about the German POW furniture and how it came to rest in our front room. At the end of the Second World War there were hundreds and thousands of German prisoners of war in this country. They weren't all sent back at once because they needed to be processed and often their labour on farms was vital to keep the production of food going. As Christmas 1945 approached my grandparents on my mother's side decided it would be the right thing to do to invite a couple of the German POWs over to spend Christmas with them. Fortunately for them (and me, as it turned out, in Chapter 10) they were given a very nice, very young German called Fritz. He had been called up when he was sixteen and luckily for him had been posted to Jersey to be one of the occupying force there. I'm sure he would be the first to admit that it wasn't the most dangerous posting for a German soldier. Nevertheless, towards the end of the war, when supplies ran short, he was forced to eat a lot of nettle soup. When we first heard this we young kids thought he'd had a very nasty war indeed.

Fritz was a very nice young man and Christmas with my grandparents passed off in the spirit so closely associated with it but rarely coinciding with it. My grandfather was a conscientious objector in the war, which was one of the reasons he was first out of the blocks forgiving and forgetting the Germans their wartime beastliness. My grandmother, on the other hand, conscientiously objected to his conscientious objection, with the ironic result that their marriage was sometimes a bit of a battleground. I often wonder what would have happened if, instead of getting Fritz, they'd had some out-and-out Nazi foisted on them for Christmas. But it didn't happen so we'll never know. Instead we began to reap the benefits of forgiveness and forgetting and those benefits came in the form of wooden furniture. The Germans POWs had a lot of time on their hands and this, combined with the fact that their camp was in a forest, meant that whittling wood had become a major industry for them. Beautiful, finely wrought carvings, inlaid marquetry and elegant pieces of furniture came out of those German prison huts. Sadly we didn't get any of these fine pieces. We got the heavy, clumsy items that could double as weapons in bar fights.

Of all the heavy and dangerous wooden items my grandparents received from the Germans, the heaviest and most dangerous was the coffee table, which had found its way, thanks to everyone else in the extended family rejecting it, into our front room in Kidlington. It was on the corner of this table that I split open my head in the middle of an epic piece of showing off. History doesn't record what exactly I was doing but no doubt it registered quite highly on the foolishness scale. Blood flowed profusely from my head (where does all that blood in the head come from?) and I was rushed to the local doctor's surgery to have some emergency stitches. In a way I consider myself to be one of the last casualties of the Second World War.

When I look back at my war wound it wasn't the most spectacular part of that particular incident. What was really surprising was the fact that my father managed to get our car started before I bled to death. My father is many things but practical engineer he is not. When he first bought the house in Kidlington, almost the first thing that happened was that the water main burst. This was in the middle of a particularly cold winter and the chances of getting a plumber out were, like the weather, less than zero – they weren't the friendly, obliging, value-for-money chappies they are today. The man who lived over the road from us must have seen my father struggling with the leak in an inappropriate fashion – I would have attempted to apply a tourniquet in that situation because I've inherited the missing practical engineer gene and I'm sure my father would have done something similar and equally ineffectual.

That's when the neighbour stepped in and introduced himself. He was the kind of man who had a very large toolbox and in it a tool for every job. He opened his box, selected the burst-main fixing tool, applied it firmly in a workmanlike fashion and the crisis was quickly over. I should imagine my father didn't know how to thank him enough. It turned out that this practical neighbour soon thought of a way for my father to thank him more than enough. His idea was that my father should purchase the car he currently had for sale at a very reasonable price. My father was looking for a cheap car so, in a moment of misplaced gratitude, he bought it. That is how it came to pass that my father became the proud owner of possibly the worst car in Britain.

To be accurate, he became the proud owner of the two worst cars in Britain because he had the undamaged front of one written-off car welded to the undamaged back of another written-off car. I say 'welded' in the loosest sense because it was actually welded in the loosest sense: the two halves of the car knew each other but weren't really that close. In order for the car to start the two halves of the car had to be forcibly reintroduced and the only way to do this was to slide under the car and belt its undercarriage with a lump hammer. This happened every time we went out as a family. We would wait patiently at the side of the car while my father, normally in his best suit, slid under the car and bashed the hell out of it. Clearly we weren't allowed to get in until my father had slid back out, in case the extra weight trapped him underneath. Had that happened, releasing him wouldn't have been the fire brigade's greatest challenge as you could have snapped the car in half like a KitKat.

The front half of the car was a white VW Vantage. Where the back half came from was anybody's guess. Shortly after my war injury I had a turning point in my life that was to absorb me completely for the next ten years until I discovered girls (and some time after that if I'm honest). My father brought me my first Matchbox car. It was a VW Vantage, which was interesting in itself because we could see what the back half of the car in the drive should have looked like. What was equally memorable was the colour, a deep purple. I soon painted it white to look even more like my father's car. Then at some stage I got it into my head to cut it in half to complete the similarity. I did this and, just like my father's car, it compromised the structural integrity of the car and it was never really much use from then on. I would like to say that I then sold the two pieces to the son of the practical neighbour but life is rarely that satisfying. Looking back, I should have asked my brother to sell it. He was seven then and was getting to the stage where he could have sold my dad's car back to our neighbour and thrown in some authentic German antique furniture just to clinch the deal.

SWIMMING NOTE: The Fatted Calf, despite his name, is a very powerful swimmer. This started when we were in Kidlington and had our first proper swimming lessons in a big pool in nearby Bicester. Well, it felt like a big pool, but when you're five years old the bath feels like a big pool. It must have been slightly bigger than a bath because I remember not being able to touch the bottom in some places. The way the instructor encouraged us youngsters to get our heads underwater was to put coins on the bottom. I'm sure today this would contravene hundreds if not thousands of Health and Safety Regulations, not to mention budgetary guidelines, but it certainly worked for me and more especially for my brother who spent the entire lesson underwater hoovering up the cash. How he didn't drown under the weight of coinage I don't know but it certainly developed his swimming muscles quickly enough. What was particularly shocking about this technique was that I distinctly remember diving for the old 3d coin, a heavy odd-shaped coin like a modern 20p coin on steroids. I don't understand why they weren't using halfpennies because 3d was almost the down payment for a house.

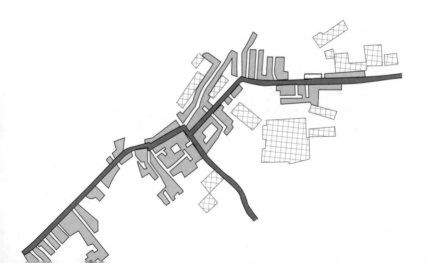

KIDWELLY, SOUTH WALES. 'HOLIDAY' LOCATION

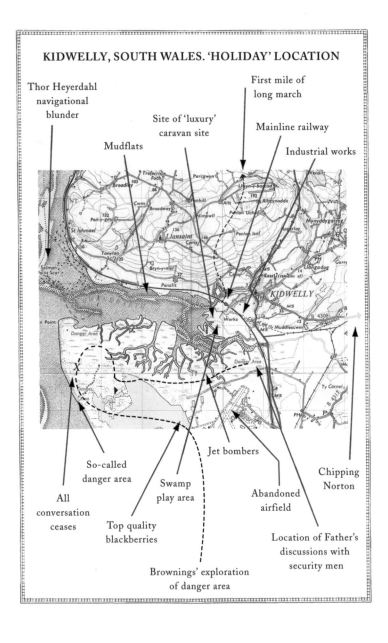

Thor Heyerdahl
navigational
blunder

First mile of
long march

Site of 'luxury'
caravan site

Mainline railway

Mudflats

Industrial works

Jet bombers

So-called
danger area

Swamp
play area

Chipping
Norton

All
conversation
ceases

Abandoned
airfield

Top quality
blackberries

Location of Father's
discussions with
security men

Brownings' exploration
of danger area

CHAPTER 5

Kidwelly, Wales

As my father travelled a lot with his work, holidays were something to be taken quickly and close to home. After some quality time spent with his maps and factoring in travel time, my father decided that South Wales would be the optimum holiday location for us as it would mean less than three hours in the car, a selection of beaches and, most importantly, interesting geographical features to study and discuss. My mother was happy with Wales as there were unlikely to be coloured lizards dropping from the ceiling. We children were also very happy (especially as my father stressed the beach aspect of the holiday and soft-pedalled the interesting geographical features angle). For a child living in Oxfordshire, Wales counts as a foreign holiday and crossing the Severn Bridge was just about the most exciting and daring way of starting a holiday short of blasting off from Cape Canaveral.

Our first holiday in Wales was on the Gower in South Wales which, because it is a peninsula, has beaches on three sides. To save money we borrowed a cottage owned by one of my father's colleagues, who was also a geographer, with the result that it was convenient for geological outcrops but remote from the beach. It's difficult to be remote from the beach on the Gower Peninsula but, as I've said, my father had a gift for finding odd places. The cottage was situated on a remote country road as far from a beach as it is possible to get on the Gower Peninsula.

The front of the house was on the road but the land behind it rose steeply in the way that much Welsh land does and, also in the Welsh manner, was liberally sprinkled with sheep. This meant that the view from the back of the cottage was more often than not a sheep's messy arse. For the Fatted Calf this was the dream holiday cottage view as he has a sense of humour that rarely emerges above the rim of the lavatory. Inside, the cottage had a homely feel if your own home had a lot of heavy hand-built furniture, which luckily ours did. The one leisure amenity the cottage had was a rope hanging from a tree with a tyre attached. Sadly, this tree was on the other side of the lane and we were absolutely forbidden from crossing it unaccompanied lest our crossing coincide with the annual rush of sheep being driven to market.

In those innocent days there was nothing for children to do on holiday apart from run around and make a racket; there were no attractions, no DVDs, no entertainment, no theme parks, absolutely nothing that encouraged children to have fun in any way. Days were spent on the beach until it was dark enough to go to bed. Unfortunately, I have very fair skin and about five minutes of sunbathing is all it takes to have me rushed off to the local hospital with third-degree burns. A beach holiday for me was therefore not short of actual physical torture and very probably constituted a serious breach of my human rights. But given the choice of a day on the beach or a day in bed with *The Puffin Quiz Book* I understandably opted for the beach. To give her credit, my mother was keenly alert to the dangers of the sun long before we had a hole in the ozone layer and to protect my fair skin on the beach I had to wear a parka with a hood which, although very fashionable, was also quite hot.

One year we went to Wales but didn't stay in our usual cottage. Someone else may have booked it before us, my father may have had a dispute with a colleague (academic disputes being the most vicious and pointless of all disputes) or he may just have been keen to broaden our horizons. It turned out that our horizons were broadened by less than twenty miles because that year we found ourselves a little way along the Welsh coast near to a small town called Kidwelly. For the first time we also found ourselves in a caravan park which was especially exciting as it seemed to imply organised leisure activities rather than geological surveys. This particular park was situated in an area of outstanding natural ugliness. One of the major sources of disfigurement was the caravan park itself which spread like a large poisonous fungus along the shoreline. About ten of the five hundred caravans had a sea view or, to be more accurate, a mud-flat view. The rest had a caravan view but at least there was the consolation of knowing that the caravan you could see in front of your own caravan, or the one in front of that one, might itself enjoy a view of the mud flats.

The whole caravan park looked like a hastily thrown together reception camp for refugees who had made the long and dangerous passage from Oxfordshire. This impression was increased by the copious use of barbed wire around the campsite and tall, unscaleable fences. In fairness some of this security was for our own safety as the mainline railway ran directly behind the caravan park. This meant that we had an almost uninterrupted view of trains passing, an added bonus as far as I was concerned, and I was very grateful to my parents for paying the extra supplement to get a caravan with railway view. In the seventies coal was Wales's largest export and long trains of coal trucks used to grind past our caravan leaving a light dusting of soot sticking to my burnt red skin.

I'd be misleading you if I said that the railway was the best bit of our Kidwelly holiday. By far the best bit was that much of the surrounding area was owned by the Ministry of Defence and was used by RAF jets as an area to practise bombing. Every now and then, the rumble of endless coal trains would be shattered by the arrival of two or three fighter bombers screaming into the estuary to launch their bombs on to targets on the far side. Often we wouldn't actually see the jets because the sea-view caravans would be in the way, but just hearing them rip past was exciting enough. The fact that the pilots resisted the temptation to wipe out the caravan park was testament to the discipline instilled by their RAF training.

Kidwelly clearly had many attractions, but long golden beaches were not among them. Given this lack of beaches, I think this must have been the year when the Long March programme was first properly introduced by my father. Long walks were an essential part of our childhood and, once my little sister became self-propelled, became the dominant part. The length of the walks was only matched by the shortness of our trousers and the shortness of our trousers was only matched by the length of our socks.

We were used to walking, of course, because everyone walked to school. In my case it was only about a mile but, because we lived on top of a hill, that mile was a one-in-four gradient. It took me about twenty minutes to walk downhill to school but, when I first started to cycle, I cut that to about forty seconds with five seconds' smoking-hot braking before I hit the school gates. The last quarter of the walk back to my house was particularly steep. When we drove back from Cubs, my friend Ian's dad used to drop me at the bottom of our hill because he wasn't sure his car would make it to the top. But, then, Ian's dad drove a Moskvich, which was a Russian car of very limited performance on the flat let alone a slope. How he managed to find, buy and then drive a Russian car is a mystery to me now but not one that concerned me much then.

The school run/walk was excellent training for our holiday walks which were designed to pass the daylight hours. Holiday walks started immediately after breakfast which meant that we became expert in massive carbo-loading before we left the table. Walks then commenced after teeth had been cleaned, shoes tied and anoraks fastened. The Parachute Regiment has worked out that five minutes' rest every hour tends to be the best balance for forced marches. This means that in a friendly competition between the Parachute Regiment and our family the Parachute Regiment would have arrived back at our holiday home approximately an hour after we did at the end of an average day.

When we were on a Long March we only rested for my father to check the map for possible interesting short cuts or further short cuts to get us back on the right track after our previous interesting short cut. We also rested for short lectures on the landscape and how it had come to look like it did. My father worked on the principle that if you understood that a cliff was composed of sedimentary limestone it would be much easier to climb. Lunch was not something we stopped for either, mainly because my father's routes and short cuts were designed to keep us close to points of geological interest and far from points of culinary or retail interest. Also there was absolutely no need to stop for lunch because we rarely ventured out without the full survival kit of a bottle of squash and a packet of Lincoln biscuits. Even now the sight of a Lincoln biscuit makes my feet ache.

Long Marches were actually a beautiful piece of child management as far as my parents were concerned: we walked at a pace fast enough to make talking not worth the extra breath so silence generally reigned; we followed tracks that only allowed single-file walking, thus further reducing the possibilities of bickering; and we walked so far that, by the time we got back, our dearest wish was to go to bed quickly and quietly with very little in the way of fuss or demands for further entertainment. Quite possibly my parents went off clubbing after we'd gone to bed and had a whale of a time into the early hours. I was fast asleep so I couldn't be sure, but I suspect that as soon as they'd had a glass of shandy and planned the following day's march, they too went to bed, lulled to sleep by the endless rumbling coal trains.

A Long March I particularly remember was the one near Kidwelly when my father decided to explore the abandoned airfield opposite us (the one that was being bombed constantly) and perhaps learn a little bit more about why it was marked as a Danger Area on the map and whether it could be usefully redesignated. The fact that there was no path marked on the map was not a discouragement to my father; it simply meant that the people who had made the map in the first place hadn't looked hard enough for one. If my father wanted to go somewhere a path would no doubt materialise or, if push came to shove, a trail could be blazed.

One of the things that remains very clear in my mind about the famous airfield walk was the tremendous number of blackberries we found. My father was good at making jam and his particular favourite was blackberry jelly, so we were always on the lookout for them. On our approach march to the Danger Area we came across incredibly thick and heavily laden blackberry bushes. I should imagine we stopped to pick some of these and, as per usual, the children would be sent into the lowest and thickest parts of the bushes to collect the berries inaccessible to adults. We were quite used to this and didn't need much encouragement to get stuck into the blackberry bushes, especially as we were wearing our thorn-proof shorts. I should imagine that at some stage while hacking through these incredibly thick blackberry bushes, we managed to penetrate the security perimeter of the Danger Area. It probably also explained why no one else had picked any of the blackberries.

Once in the Danger Area the going became a little easier. We came across dilapidated red-brick buildings that must once have been the living accommodation on the abandoned airfield. Soon the thorns and blackberries closed in again and my father took the view that by far the most sensible way forward was to walk along the main runway. My brother and I were old enough to read maps by that stage, especially relatively easy words like Danger and Area. It also began to dawn on us that if we could see the caravan park over on the other side of the estuary then surely we must be in the area we could see across the estuary from the caravan park, which was where the jets dropped their bombs.

Often the family would walk in silence because we were short of breath and near exhaustion. This time we walked in total silence to pick up the sound of approaching jet bombers. My father, who had started to explain to my mother that Danger Area didn't necessarily imply an area of danger, also fell silent as I believe he was concentrating very hard on navigation and any cuts that would be genuinely shorter and possibly safer (without, of course, admitting that we were at any time lost or in any kind of danger). As it turned out we weren't bombed or even lightly strafed; I think I would have remembered that. I do remember arriving at a security checkpoint from the dangerous side and my father being involved in long and animated conversation with the men on the gate. My mother held us back in case of unexploded expletives and after an exchange of opinion with the security men we were let through into the Safety Area beyond. I expect the security men were pleased to see my father as he would have explained many ways in which the security of the airfield could be improved and he probably threw in some tips for good blackberrying.

On the positive side, after that walk we probably had blackberry and apple crumble for tea. All of our holidays as children were self-catering; asking someone else to cater for you was, to my mother, tantamount to financial suicide and an open invitation to wipe out the entire family through food poisoning. Before we left home for our holidays, my mother would generally pack enough food for the entire week in the sure knowledge that food would be hard to come by in Wales and what little there was would have to be killed and butchered by hand.

That's why I'm still surprised that one year we had a holiday in a half-board hotel. The Fatted Calf and I assumed that the half-board referred to our sleeping arrangements and we wondered whether it was a half-board each or whether we'd have to share the half-board and end up with a quarter of a board each. But, miracle of miracles, half-board actually meant that we had a cooked breakfast and a cooked dinner separated only by the obligatory thirty-mile walk. In the little dining room at the end of the day our family attracted strange looks from the other guests as we ate every single scrap of food in total, silent concentration. At one stage the hotel proprietor felt moved to ask us whether he was actually providing enough food. My father assured him that he was in a manner that implied he clearly wasn't and I believe extra slices of bread were provided by the management the following day.

There is no circumstance in life, however dispiriting, degrading and dehumanising, that doesn't also provide a commercial opportunity and so it proved in our caravan park at Kidwelly where the Fatted Calf pulled off one of his first and most impressive marketing coups. The two main locations for social activity on any kind of campsite are the facility where you empty your chemical toilet and the camp shop, generally located alarmingly close to each other. I am relieved to tell you that the Fatted Calf's amazing marketing coup took place in the shop rather than the other hub of social activity. One of the things they had for sale apart from dinky packets of Weetabix and single sachets of sugar was a selection of postcards. Some of these were panoramic views of the caravan park that some people would presumably want to send to their relatives in prison to cheer them up. Sensibly, the management had made sure the vast majority of the postcards were of the naughty seaside variety featuring buxom women and puny men doing something dangerously vulnerable to innuendo. These cards may be viewed as totally innocent now but to us young boys they were the cutting edge of filth.

The Fatted Calf knew that sex sells even before he really knew what sex was. Somehow he managed to buy the shop's entire stock of postcards. The venture capital would have come from vending machine slots, his pocket money and my pocket money. I didn't voluntarily co-fund this venture: it's just that my pocket money was extracted from me week by week by my brother selling me individual Opal Fruits at hugely inflated prices with the promise that the selection would always be pre-sorted to filter out the vile strawberry ones. After the summer holiday, he returned to big school and sold the postcards to his classmates at a huge mark-up (it wouldn't surprise me if he'd also sold a job lot to Mr Franklyn, the short-of-breath carpentry teacher). It's interesting that my parents didn't stop him buying the cards although I think they probably would have done if they'd thought he was just buying a couple for his own personal gratification. Maybe they were just relieved that we weren't going to send any postcards of the caravan park to anyone we knew.

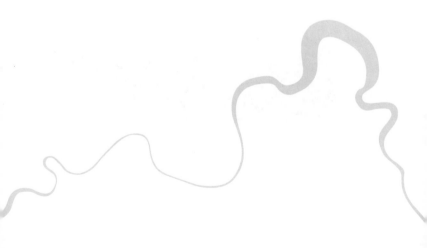

MAP OF JAMAICA TAKEN FROM
SUPERB SOUVENIR ASHTRAY

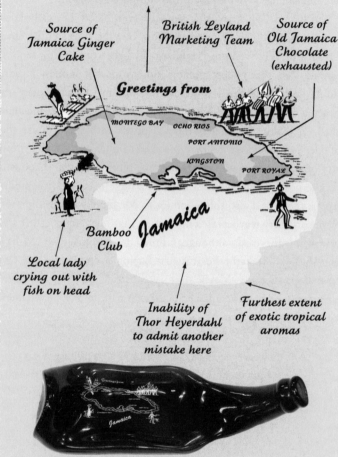

Closest Nina and Frederik
got to Jamaica (6000 miles)

Source of
Jamaica Ginger
Cake

British Leyland
Marketing Team

Source of
Old Jamaica
Chocolate
(exhausted)

Greetings from

MONTEGO BAY   OCHO RIOS

PORT ANTONIO

KINGSTON

PORT ROYAL

Bamboo   *Jamaica*
Club

Local lady
crying out with
fish on head

Inability of
Thor Heyerdahl
to admit another
mistake here

Furthest extent
of exotic tropical
aromas

CHAPTER 6

Kingston, Jamaica

I've never actually been to Jamaica and, quite honestly, I don't think I ever need to go as my mental picture of Jamaica is pretty much complete. I also know that the picture I have is pretty much completely wrong and going there would involve such a total rebuilding of my mental image of the place that it's just not worth the effort. I haven't been to Russia either but I know the picture I have of it in my head – wide-open spaces, ugly buildings, bad roads, organised crime, chess, high cheekbones and vodka – is substantially correct. Jamaica, however, I've got very, very wrong.

My first image of Jamaica came from my father who visited the island while he was in the navy and his ship was doing a tour of the Caribbean. He always used to reminisce fondly of Kingston, Jamaica's capital city and popular port of call for the navy. 'Ah yes, the Bamboo Club,' he would say, looking fondly into the middle distance. 'You could smell the syphilis ten miles offshore.' I was very young when I first heard him say this and I imagined that syphilis was some kind of fragrant tropical flower like the bougainvillaea and that the Bamboo Club was an elegant colonial drinking establishment set among this luxuriant vegetation. My impression changed about ten years after first hearing this because I inadvertently stumbled upon sexual education.

I'm not quite sure when or if I developed sexual understanding but I can say with absolute certainty that I knew nothing about the reproductive act when I was eleven years old. To be honest with you,

I now have three children and still know very little about it. I had a beautifully sheltered upbringing and my introduction to the big wide world of sex came in the most inappropriate and potentially embarrassing way.

It was a rainy weekend afternoon and my father was doing what all fathers need to do on a regular basis – forming a deep and satisfying bond with his newspaper. We children had been trained to be silent and still when he was having intimate moments with his newspaper and we knew that the best way of achieving this was for us to read another newspaper he'd previously had a relationship with but which was now over. That's how I came to be scanning the Personal Ads in the local free newspaper and how I came to ask the rather challenging question, 'Dad, what is a VD clinic?'

I realise now that the question is probably not the best departure point for a gentle introduction into the subject of loving sexual relations. The newspaper my father was currently reading can't have been terribly interesting that day because he decided to put it down and explain to me the facts of life. I remember nothing of what I was subsequently told but I do have a very clear impression of the pattern of the curtains we had in the living room at that time because that's what I was staring at with incredible intensity in order to avoid catching my father's eye as he reeled off various totally unbelievable and unimaginable anatomical scenarios. Clearly when your starting point is the VD clinic you have to do some serious backtracking before you even get to the birds, let alone the bees. It must have been one of the longest three hours in my father's life.

Mind you, he must have felt he'd covered the ground pretty comprehensively because sex didn't rear its ugly head in our house for about five years after that lecture. Then, one day, my father came home from work, plonked a very nondescript hardback book on the kitchen table and said, 'Who wants to read a sex manual?' Of course by this time I was up to my spotty neck in puberty and I knew that I knew practically nothing about sex so I naturally pretended that there was nothing left for me to learn on the subject. The Fatted Calf, being one year older, knew slightly more so had to pretend slightly less, but we both immediately said 'no thank you' as if our palates for that kind of thing were jaded beyond measure.

In the next twenty-four hours we'd read the thing from cover to cover and memorised large chunks and discussed whether some of the later chapters might actually be totally made up. The book stayed in the house for the next couple of decades and proved a valuable reference work, especially for my mother. On reflection, I think she was the most shocked of all of us at the contents of the book as it went well beyond matters covered in the problem pages of *Good Housekeeping*.

For me, Kingston, Jamaica, would simply be the home of the Bamboo Club and source of venereal unpleasantness were it not for the fact that I had another impression of the island stamped hard and deep on me from the very opposite end of the pleasantness spectrum, courtesy of the Danish calypso sensations Nina and Frederik. As a way of immersing themselves in Caribbean and Latin American culture before they left for El Salvador, my parents had invested in a record of Nina and Frederik singing calypso-style songs that were currently being popularised by Harry Belafonte. I can only assume that Nina and Frederik were on a parallel mission to make them less popular.

Nina and Frederik were a perfect blond couple from Denmark. In a well-ordered universe they would have been spokesman and woman for the Danish Dairy Board. Frederik had a neat, clipped beard which left him halfway between a knitwear model and a U-boat captain. His wife looked as if she'd just scrubbed herself clean on an outdoor washboard. Together they were so wholesome you could have made healthy breakfast bars out of their droppings. Unbelievably, he was also a baron and she a baroness, both of ancient aristocratic lineage. Despite these overwhelming cultural handicaps they had bravely chosen to reinterpret Caribbean music for a Scandinavian and North European audience. Part of this extraordinary process was attempting a local Caribbean accent. The result came out halfway between Denmark and Barbados, which placed them in mid-Atlantic, and they did indeed sound as if they were on the point of drowning. I'm not sure why I'm being so rude about them because I loved their music, as did the vast majority of Scandinavia and North Europe.[*]

'Jamaica Farewell' was one of their big hits. I know every word and still sing it in my bath at maximum volume (it's my attempt to popularise calypso music in the neighbourhood where I live). The lyrics of this song gave me a whole different impression of Jamaica:

'Jamaica Farewell'
by Erving Burgess

Down the way where the nights are gay
And the sun shines daily on the mountain top
I took a trip on a sailing ship
And when I reached Jamaica I made a stop

CHORUS:
But I'm sad to say, I'm on my way
Won't be back for many a day
My heart is down, my head is turning around
I had to leave a little girl in Kingston town

[*] *No West Indian I've met has ever heard of Nina and Frederik.*

KINGSTON, JAMAICA

Sounds of laughter everywhere
And the dancing girls swaying to and fro
I must declare that my heart is there
Though I've been from Maine to Mexico

CHORUS

Down at the market you can hear
Ladies cry out while on their head they bear
Ackie rice and salt fish is nice
And the rum is good any time of year

CHORUS

There have been times when I've wondered whether the girl in question was one he'd met in the Bamboo Club and the fact that his heart was down and his head was turning around might perhaps have been the first early symptoms of something venereal he'd picked up there after a bit of swaying to and fro. But I'm sure Nina and Frederik simply wouldn't have covered the song if there had been any kind of unwholesome implication. Another song on the album was called 'Limbo' and it was about a man who wanted someone 'limbo, limbo like me'. I'm pretty sure this was shorthand for someone who was remarkably accommodating and athletic in bed but the way Nina and Frederik sung the song it sounded more like a call for volunteers for school sports day.

The lines 'Down the way where the nights are gay/[Bamboo Club reference, surely] And the sun shines daily on the mountain top' confused my mental image slightly because, while I could picture the sun shining on the mountain top, it did plunge the rest of the island into darkness. The other line I struggled with was 'Ladies crying out while on their heads they bear'. I didn't pick up that these were probably trade cries such as 'Lovely fish'. I thought instead these must be cries of anguish along the lines of 'Help, there's a fish on my head', a fish that had somehow found its way there in the darkness.

There was one other thing that completed my total misunderstanding of Jamaica, its people and culture and that was an ashtray my father brought back from a later trip to Jamaica. I would love to say this was from the Bamboo Club but it wasn't and I can guarantee that I know exactly where it came from even though my father would never admit it. At every airport there are shops that sell souvenirs for travellers to take home with them. The stock never varies from Mumbai to Medellin or, indeed, from Maine to Mexico, and includes T-shirts, watches imported from Switzerland, oversized Toblerones (also from Switzerland, strangely), Swiss Army knives (made in China), laptops unusable outside the country of departure, glossy books featuring all the country's landmarks in unnaturally bright sunshine, and dolls in national costume. One other thing you could guarantee to find in the gift shop was an ashtray of local origin, useful for the flight home where everybody smoked like a chimney.

This 'Greetings from Jamaica' ashtray was in the shape of a flattened brown beer bottle. Decorating the flat middle portion was a handpainted map of Jamaica featuring steel bands, ladies crying out while on their heads they bear, and exotic sounding places like Montego Bay. I would love to be able to tell you that the Bamboo Club was marked on the map but it wasn't. Even without the Bamboo Club I always thought there was something slightly odd about this tropical paradise being depicted on a squashed beer bottle that served as an ashtray.

On further reflection I think what's equally odd is that in the darkest days of British Leyland they decided to call one of their least inspiring, most unmemorable and doggedly dull cars the Austin Montego. Somehow the marketing men of the time must have thought that summoning up the spirit of Jamaica would entice people to go for this model rather than the cheaper, better looking, more reliable German equivalent. But as the Montego drove like a hammock, rusted like shipwreck and was little more than a glorified ashtray itself, maybe the marketing boys were just being honest for once.

What the ashtray also tells me is that they have no national costume in Jamaica because, if they did, a doll wearing a miniature version of it would have been brought home by my father to clog up our house along with eighty other dolls in national costumes rivalling each other for colourfulness and sheer lack of practicality. The expression on the faces of these dolls never varied: it was always a bemused, slightly shocked look, as if to say, 'Why am I wearing this ridiculous outfit?' It's good to know that when you leave England there are no dolls in our national costume, so that children in Jamaica don't have a bemused looking Morris Dancer on their shelf blighting their childhood and misrepresenting our country. At least when you get a double-decker bus or a London taxi back to Kingston you can have a nice game of bus lanes and congestion charges. No child plays with dolls in national costume except to line them up for Barbie and Action Man to make fun of for being badly dressed, sad-looking losers.

A few years ago I used to live in a very nice village in Oxford-shire with the lovely name Kingston Bagpuize. People generally remember the village because of its fantastic name, which they mostly associate with the mangy pink and white striped cat Bagpuss. Happily, there's no relation between the two. Kingston was called plain Kingston until the invasion of England by the Normans in 1066 under William the Bastard. Once the Bastard had the English under control, he started parcelling out the country to the knights he had brought with him. One of these was Ralf de Bachepuise, from the village of Bachepuise in Normandy, and he got Kingston as a reward for doing something in the Conquest (clearly not a great deal otherwise he could have got somewhere a bit more impressive for his troubles). That's how the village became Kingston Bagpuize and there's still an undercurrent of bitterness towards the Normans in some of the outlying parts of the village.

Like all villages in England, Kingston Bagpuize has a big sign at each end with the name of the village on it. Most village signs also say 'Please Drive Slowly' but for some reason ours said 'Put Your Foot Down Lads!' On a more truthful note, we did for a while have a very interesting village sign. The early morning commuters who slipped out of the village just after dawn to Oxford, Newbury and even London, found on their return one dull winter day that our sign, instead of reading 'KINGSTON BAGPUIZE', now read 'KINGSTON JAMAICA'. It was no rushed piece of graffito either: the lettering was exactly the same as the official Highways Agency lettering and, especially in the pale winter light, it was impossible to see the join. For the few weeks the new sign remained up, a little bit of Caribbean sunshine crept into our village as only the hardest of heart or shortest of sight could have passed it without smiling.

No one ever found out who changed our sign and it wasn't for lack of trying either. There were certain villagers who made it their business to know everything. They acted as a form of rural search engine: you just typed in what you wanted to know about (or mentioned it in passing) and you were immediately offered up all sorts of information, much of it useless but some absolutely spot on. But no one knew anything. Whoever it was I admire them, especially the care they put in to making the sign indistinguishable from the real one. I hope he or she has gone on to bigger and better projects. Maybe they've flown to Jamaica and changed the airport signs to 'WELCOME TO KINGSTON BAGPUIZE'. I don't know because I've never been there but maybe I should go and check one day and, while I'm there, have a look at the real Jamaica.

WILLIAM TURNER'S VIEW OF OXFORD PAINTED FROM RALEIGH PARK WHERE I COLLECTED DUNG

North Oxford home of terror-women

Steam rising from heated academic discussions in Oxford

Oxford in Oxfordshire

Old Father Thames (rolling along)

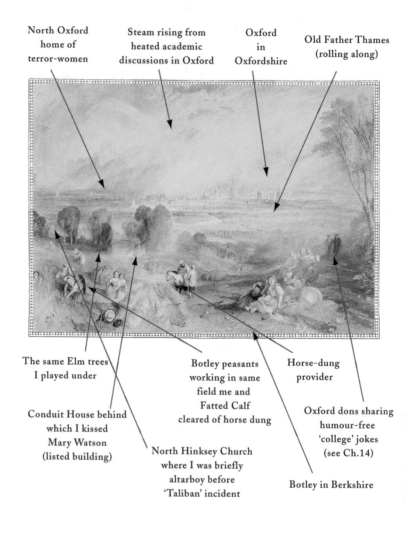

The same Elm trees I played under

Conduit House behind which I kissed Mary Watson (listed building)

Botley peasants working in same field me and Fatted Calf cleared of horse dung

North Hinksey Church where I was briefly altarboy before 'Taliban' incident

Horse-dung provider

Oxford dons sharing humour-free 'college' jokes (see Ch.14)

Botley in Berkshire

CHAPTER 7

Botley, Oxford

If people ask me where I'm from I generally say Oxford. That gives the impression that I'm sophisticated, educated and cultured. In fact, if I'm honest, I'm from Botley, which is three miles west of Oxford and a precipitous drop down the sophistication scale. There's no way you can say the word 'Botley' and sound sophisticated, educated or cultured. I've heard people say 'Boatley' to make it sound a little bit more like a waterside development. But Botley means the bottom field and Boatley would mean the boat in the field, which wouldn't make any sense at all. People from Botley who are worried about appearing common usually say they are from North Hinksey, which sounds a lot better. It's only a stone's throw from Botley, but you have to throw the stone right across the A34, the main trunk road between Southampton and the Midlands which now separates the two villages as effectively as the Berlin Wall.

On reflection, Boatley might make a little bit of sense because, at the bottom of Botley (Botbotley), there is a little babbling brook called the Seacourt Stream. The Thames splits above Oxford in a sudden loss of confidence and slips through the city in a series of small offshoots, the smallest and most westerly of which is the Seacourt Stream. In medieval times there was a village called Seacourt where fish used to be caught to feed the monks and nuns that littered the place. If developers were building a new residential estate in this area now (it's smack in the middle of the flood plain so they're probably queuing up) they would call it Seacourt Reach rather than Botley

Villas because they have an instinctive feel for what sounds like an exclusive premium development and what does not.

This little Seacourt Stream punches above its weight, or flows above its channel, because for about a thousand years it formed the western boundary between the Royal County of Berkshire and plain old Oxfordshire. Botley was in Berkshire, and Oxford, three miles away over the Seacourt Stream, was in Oxfordshire. I suppose it was fortunate that the stream didn't gradually shift its course to the east in those thousand years otherwise Oxford would have found itself in Berkshire and that would have left Oxfordshire looking pretty silly.

In 1970 we moved the seven miles from Kidlington in Oxford-shire to Botley in Berkshire. Then, a mere four years later, we moved back from Berkshire to Oxfordshire. Although *we* didn't actually move, Oxfordshire did. Some flared-trousered, overzealous, jumped-up, pettifogging, desk-bound bureaucratic vandal thought one idle afternoon he'd bin a thousand years of history and wrench the Vale of the White Horse and several beautiful market towns (Wantage, Abingdon, Faringdon) from the bosom of Berkshire and give it to Oxfordshire as a kind of buffer state because Oxford was too close to the border. Botley is now Oxford's equivalent of the Golan Heights: it doesn't really belong to Oxford but is kept in place by sheer weight of local government bureaucracy.

The reason Botley was called the bottom field is because it's at the bottom of Hinksey Hill. This is one of the hills surrounding Oxford from which you get the lovely views of the dreaming spires. For a long time Botley was a few labourers' cottages at the bottom of the hill and stretching away from it were two roads that led straight back up the hill like a big V sign to the labourers beneath. These two roads were Harwell Hill and Hinksey Hill and this is where the gentry lived. Then one day the man who owned the whole hill decided to let down everyone on the two posh roads very badly and sold the rest of the hill for a huge estate to be built on. This massive estate was built on the site of Sweatman's Farm. You won't be surprised to know that the developers called the estate Elms Rise.

When we moved to Botley the dominant feature (apart from the remarkable Mr Lancaster*) were the elm trees. They were everywhere. The estate was called Elms Rise, the shops were called Elms Parade, my infant school was called Elms Road Primary School and there was even a music teacher at the school called Mr Elms, although I'm sure that was just a coincidence. Thanks to Dutch elm disease† they were wiped out almost overnight yet their name lives on everywhere around Botley. It's a little bit like North Hinksey over on the other side of the A34. It was named after the fabled Saxon warrior Hengist, partner of Horsa, who was very popular at the time of the village's foundation some time in the eighth century AD. There were a lot of places named after him: around Oxford there is North Hinksey, South Hinksey, Hinksey Hill, New Hinksey and Ferry Hinksey. In terms of easy naming, Hengist was the Nelson Mandela of his day. Now, like the Elms, old Hengist doesn't really make much sense to the locals any more although one of my friend's rabbits was called Hengist for the short time he lived before dying of shame.

As I mentioned, the parade of shops in Botley was called Elms Parade and it contrived to combine the worst of fifties architecture at the front with the worst of sixties architecture at the little precinct at the back. If any word in the English language smells of urine it's the word precinct. When the architects of this little precinct sold it to the local planners they would have used words in their presentation like Manhattan, New Horizons, Liberation, Freedom, Expression. I'm sure

* See Chapter 18 for more on the legendary Mr Lancaster.
† Called German elm disease in Holland, Belgian elm disease in France, Russian elm disease in Poland and Great Satan elm disease in Iran.

the words Skateboard, Dog Shit, Graffiti, To Let, Concrete, Faceless, Dispiriting and Grim wouldn't have made it in their presentation but that's how Elms Parade presented itself to the world for many years.

Then, in the nineties, when people began to wake up to the fact that sixties architecture was a bit of mid-life crisis that did no one any good, they decided that the whole place needed sprucing up. The local architects would have explained to the local planners that the best way of doing this would be to erect Mediterranean-style covered walkways which would then have plants growing from them to give a Hanging Gardens of Babylon feel in the space between Iceland and the chip shop. To complete the sun-drenched piazza feel a lick of magnolia paint was also given to the entire parade, including the big end wall opposite Barclays Bank. When the whole place was finished, the locals decided to express their appreciation for the social and environmental uplift by painting all across the big end wall *Botley Boot Boys Kick to Kill*. This graffito stayed in place for a good few years as the entire budget for the beautification of Elms Parade had been spent on the Hanging Gardens.

There were two interesting things about this bit of graffiti. The first was that no one ever bothered to write anything else on the wall even though there was plenty of room. Maybe because everything had been said already. Or maybe because the Boot Boys in question had made it known that they didn't want anybody invading their artistic space. The second most unlikely thing about this graffito was that it was perpetrated in rather elegant cursive script. This gave the very unusual impression that the Botley Boot Boys were a highly educated group of intellectuals who had renounced their academic studies for a spot of gratuitous violence and vandalism. There's every chance that it was the work of Oxford University students lodging in the area who have probably now grown up to be hard-core accountants giving their clients' accounts a good kicking.

On moving from Kidlington to Botley, we expanded from a three-bedroom house into a four-bedroom house. It was through this subtle architectural variation that I first became aware of my sister. When we first arrived in the house on moving day, the Fatted Calf and I thought the best way we could help with the complicated logistics of the move would be to tear around the house shouting. Our first self-appointed task was to explore all the rooms, shout in every one of them and then claim one for our bedroom. It goes without saying that the Fatted Calf claimed the largest bedroom with the best outlook, safe in the knowledge that our parents would be happier in one of the smaller rooms at the front. Getting him out of that room was like getting the Russians out of East Germany and added to the joy of moving house for my parents. In the lengthy process of negotiating for the bedrooms, my sister was allocated a room. Thus for the first time we had to acknowledge the fact that she had started to physically impinge on our lives.

My little sister is four years younger than me and thirty years more mature. The effect of my sister on our family has always been similar to waking up on the morning after a heavy snowfall. Even when you're still in bed you can feel instinctively the arrival of a great calmness and stillness outside. When you fling open the curtains, you see that the landscape, however mundane, has been transfigured into a vista of great beauty. Such is the effect of my sister who, for obvious reasons, is generally known as the Sainted One.

Appropriately enough, the Sainted One arrived in a heavy snowfall. It was the winter of 1968, generally considered a cold one by people who make it their business to remember weather patterns from year to year. My mother was staying with her mother in Loughton, Essex, and when the contractions started my father drove to Epping hospital through the rapidly thickening snow. After a few hours nothing much had happened (my sister is never in a rush to do anything and obviously prefers baths to showers) so the hospital sent my parents away. The snow was now so heavy that they decided not to go back to my grandmother's but to go to the local cinema instead.

The film showing on that day was *Poor Cow*, the opening scene of which was the most visceral, graphic, technicolour portrayal of a woman screaming in agony during childbirth. Shortly after this opening scene, my mother for some reason felt the need to get back to the hospital. Unfortunately, while they'd been watching the training video in the cinema, the local population had been injuring themselves in vast numbers in the snow and ice and filling up every available bed in the hospital. By the time my mother was guided through to the delivery room there was only one very junior doctor on hand to oversee matters. He asked my father in a half-joking way whether he'd done this sort of thing before. As you'll remember my father is genetically incapable of admitting to any kind of ignorance or gap in his knowledge so would no doubt have given an answer that implied he was a lay member of the Royal College of Surgeons.

About an hour later it became apparent that neither my father nor the junior doctor was helping in any way (at one stage I think my father succeeded in getting the junior doctor to stop breathing). Fortunately a midwife arrived, sent the junior doctor packing and my father out to help the receptionists improve their job efficiency and calmly helped my mother deliver our little sister. After a brief inspection to see whether his third child was a girl or a boy and that it and his wife were OK, my father went home. The snow kept falling and the locals, if they hadn't already injured themselves, decided it was best to stay in. Pretty soon, the hospital grew quiet and for the next five days my little sister, snug and serene in her swaddling bands, presided over a calm and peaceful hospital muffled in deep, undisturbed snow and caught in the crepuscular light of weak winter sun.

When she eventually came home, my sister had the same calming effect. Even today when she enters a room it's like a bishop entering a cathedral: everyone feels an instinctive need to stand up or kneel and generally confess their sins and try harder to be good.

Theoretically, having a four bedroom house meant that we had one bedroom each (the parents being one flesh). In reality it didn't work out like this because my father was simultaneously running a boarding service for international outcasts. Oxford University has traditionally been the home of lost causes and my father had extended this concept by turning our house into a kind of Battersea Underdogs Home. Anybody who was having a hard time in the world for geopolitical reasons was welcome to come to Botley and have one of our bedrooms. The standard occupant of my bedroom wasn't me but was generally some kind of Third World freedom fighter who came and went depending on the success of the struggle at home. The freedom fighters tended to blur into one angry bearded mass, but we did have two other unforgettable boarders.

The first was a young Spanish priest by the name of José, a quiet and transparently holy man who quite possibly was the victim of the last years of Franco's Fascist dictatorship in Spain. On the morning of the day José arrived, my father had delivered to the house a ton of horse manure which he was planning to use on his potato patch at the bottom of the garden. As there was only a narrow side passage from the front garden to the back garden, this load of dung was tipped out on to our front drive directly opposite the front door ready for wheelbarrowing away.

José was a mild mannered and cloistered individual from Valladolid who had not travelled to England before. He arrived at Oxford station to be met by father and was then whisked back to our house, probably via my grandmother, who would no doubt have welcomed him in her curlers. My father would have taken the traditional tapioca pudding and José would probably have had to hold this on the journey back to Botley wondering what it was, what was going to happen to it and whether he'd be involved in any way. Then he would have arrived at his lodgings in Botley to be greeted by a pile of horse dung higher than his head gently steaming in front of the house. My father would have felt absolutely no need to explain this and José would have assumed that this was another traditional English custom and not to be remarked on. He may have crossed himself and said a silent prayer but that's probably just embellishment on my part.

As it happened, the next day saw some of our other visitors pop by. They were anti-apartheid journalists who had been expelled from South Africa and who were spending their time seeing Europe from the back of their camper van. When they dropped by our house, they said what they wanted more than anything else was a bath. This wouldn't have been so bad had not the very next night another South African friend flown in from New York and said that what he really needed most in the world was a bath. He disappeared upstairs and had one and by now José thought that the polite thing to do on visiting an English household was immediately to ask for a bath, possibly to rid the person of the smell of horse dung acquired on entering the premises. Interestingly, José was not noticed to bathe, shower or indeed wash for his entire stay, probably because he thought he would catch the mad English disease.

Another interesting guest was an exile from Zimbabwe, or Rhodesia as it then was. He was a poor black farmer who'd fallen foul of a scam by which con men promised anyone a free education in England if they bought their flight and paid a hefty 'arrangement' fee. Sadly, this was a total fabrication and our friend Langton Machoko was one of the poor Rhodesians who found themselves at Heathrow, totally unexpected, abandoned, minus their life savings and with no one having a clue what to do with them. Fortunately for Langton, my father can sniff out an underdog within a hundred-mile radius and he soon found himself on the road to Botley with a tapioca pudding in his lap.

Langton also arrived at our house at a very auspicious time. My father had bought a giant antique desk for his study and was trying to haul it upstairs with the aid of three other Oxford academics. Now these four Oxford academics probably had twenty degrees between them and no doubt a smattering of Nobel prizes but they couldn't shift furniture for toffee. Langton, being a practical, common-sense kind of guy, suggested a few sensible angles and the desk shot upstairs. Within hours of his arrival he'd proved that he had more savvy than the cream of Oxford intelligentsia. It wasn't the education he expected but he certainly enjoyed it and reminded my father of his success about seventy-five times more than my father needed to hear about it. We also saw Langton crying shortly afterwards when we went to Oxford to show him the sights. We drove past some men digging the road and Langton was soon wiping silent tears from his eyes. When my father asked why, he said he couldn't believe he was a black man sitting inside a car, watching white men digging up the road.

Langton was an affable type and we kids very quickly learned to love him. The only time he looked outfaced by this strange new culture he'd been abandoned in was when my parents asked him whether he'd like to go swimming. He suddenly looked so worried that my mother asked him whether he could actually swim. Langton said that swimming wasn't the problem but that at home they used to throw a goat in the river to keep the crocodiles busy while they were splashing about. He hadn't seen a goat in his stay in England and wondered how we'd manage that problem. The Fatted Calf and I swiftly reassured him that this was precisely what the Sainted One was for and off we went.

Langton eventually went to Warwick University and did very well, but then his life took a turn for the worse. He returned to Zimbabwe, which Robert Mugabe was in the process of systematically destroying and, to make matters really desperate, he decided that he would set up a Rover dealership. Sadly we never heard from him again although we used to get regular Christmas cards from some of the other run-of-the mill freedom fighters, sent from jail or from the presidential palace depending on how they'd got on.

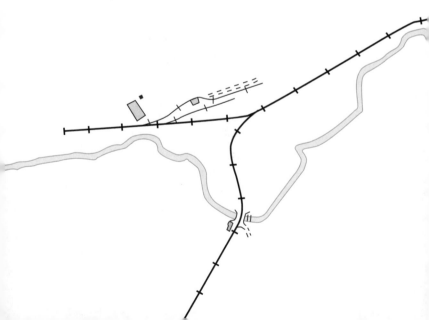

LOUGHTON, ESSEX - PARENTS' BREEDING GROUND

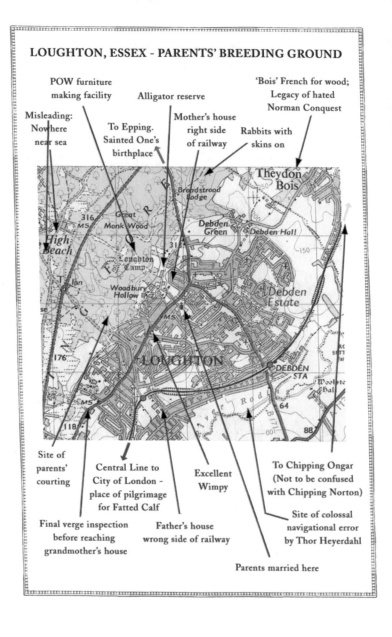

POW furniture making facility

Alligator reserve

'Bois' French for wood; Legacy of hated Norman Conquest

Misleading: Nowhere near sea

To Epping. Sainted One's birthplace

Mother's house right side of railway

Rabbits with skins on

Site of parents' courting

Central Line to City of London - place of pilgrimage for Fatted Calf

Excellent Wimpy

To Chipping Ongar (Not to be confused with Chipping Norton)

Final verge inspection before reaching grandmother's house

Father's house wrong side of railway

Site of colossal navigational error by Thor Heyerdahl

Parents married here

CHAPTER 8

Loughton, Essex

Loughton in Essex is where London cab drivers go to die (as long as they can get a fare going in that direction). It's also where both my parents were born. I grew up with a very clear mental map of Loughton. The far reaches of the Central Line run through it, like a red aorta carrying the life-giving oxygen of Essex girls into the heart of the City of London. On one side of the railway is the right side of the tracks, which are the leafy slopes leading up to Epping Forest, and on the other side of the railway is the wrong side of the tracks leading down to the marshy ground of the Roding Valley. My father was from the wrong side of the tracks and my mother was from the right side. My mother, never normally interested in railways, has never let him forget it. At the top of the right side was York Hill, which is where they sat on a bench and courted. They then went back down to Loughton High Road and got married in St Mary's Church, not on the same day although that wasn't unusual in those times if the courting got a bit out of hand.

In the normal run of things my grandmother's house should have been on the wrong side of the tracks given the start she'd had in life. My grandmother is the mother of all Essex girls. She was born and bred in Essex and embodies all the virtues of the great tribe of Essex warrior women that drive the City of London forward as the world's premier financial centre. As far as I'm aware she has none of the vices associated with Essex girls and, even if she did, I'm not about to mention them since she's still with us. As I write this she is 102 and beginning to contemplate early retirement. However, as long as she can get someone to bring her the local newspaper with the property pages, she will be dreaming up new property development schemes. It wouldn't take a molecular scientist to work out that she is the source of a lot of the genetic material that makes up the Fatted Calf.

My grandmother was born in a tiny village called Kelveden Hatch in a remote part of the beautiful Essex countryside that no one knows about (you won't like it so don't go there). There are only a few small cottages in Kelveden Hatch so it's remarkably unremarkable. However, from about 1950 the village became officially unremarkable at the very highest level. At the height of the Cold War (a war either personally instituted by Dr Kissinger in another of his hare-brained schemes or brought to a close by his brilliant diplomacy – one of the two) Britain decided it needed a huge bunker to house the Prime Minister and all the essential functions of government in case the Russians decided to nuke London in a gesture of socialist solidarity with the Labour Government.

A huge underground complex beneath Kelveden Hatch was built and they put the top-secret entrance to it under a small bungalow. The only way you could gain entrance to the bunker was by knocking on the front door as if you were nipping round to borrow some sugar. I think for security they had some absolute dragon living in the bungalow such as my Aunty Rose, who you wouldn't visit under any circumstances short of an all-out nuclear war. Nowadays the bunker is a tourist attraction with signs everywhere pointing to 'Secret Nuclear Bunker'. I'm hoping that these signs weren't up during the Cold War itself.

I don't know too much about my grandmother's early life apart from the fact that she went to school on stilts because, more often than not, the road was flooded. You hear a lot about Ethiopian and

Kenyan children running ten miles to school everyday and growing up to become superb athletes. If there were a stilt race in the Olympics (and why isn't there, if you can have beach volleyball and throwing a hammer, for goodness' sake?) my grandmother would have certainly been among the medals. This would have been at the Berlin Olympics of 1936 and she could have shared the limelight with the immortal Jesse Owens and helped him show the Nazis a thing or two about great sporting/stilt walking ability.[*]

My grandmother was the oldest of thirteen children. That was the norm in the early part of the last century and if you only had six or seven you were assumed to be practising some kind of rigid system of family planning. I think my mother was born before the last of her aunts and uncles, so it all got pretty confusing. My grandmother grew up looking after all her brothers and sisters and then she married a man (the conscientious objector) who also required looking after. After the First World War there was a great shortage of men and I'm pretty sure my grandmother accepted the first one who turned up who could read and write, was relatively clean and showed an interest. Compared to looking after twelve siblings at her mother's home, looking after one mild-mannered man in her own home was almost like a rest cure.

However, my grandmother wouldn't know how to rest if her life depended on it (that's probably why she's still alive), and she certainly wasn't about to rest once she got married. Instead she became an agent for a corset company called Spirella. If you're a woman between

[*] Is there a link between throwing the discus, which was one of the original sports in the ancient Greek Olympics, and the modern Greek tradition of throwing plates? And are the Scandinavians good at throwing the hammer because they used to worship Thor, the god of thunder, who used to do exactly that? And are the British good at middle-distance running because we don't normally have to go much further before we've got where we want to go?

eighty and ninety reading this, you'll probably have just experienced a sharp pain around your ribcage. Wearing a Spirella corset was very similar to squeezing your entire body into the cardboard inner tube of a toilet roll. The advanced Victorian engineering of the Spirella corset meant that it was able to apply pressure at the equivalent pounds per square inch of a hydraulic ram. Beautifully rounded women up and down the country queued up to inflict the Spirella corset on themselves. It was the 1930s and the new, liberated, curveless flapper look was in vogue and, if the liberated look meant 500 lb of pressure per square inch, so be it.

Spirella manufactured the original whalebone corsets. Without in any way wanting to be disparaging about whales, they are mostly blubber and their bones do a terrific job holding all that blubber in. The bones of whales were therefore conscripted to do the same job for the shapelier women of inter-war Britain. It must have been a great relief to the whale population when the human population discovered Lycra. Thankfully, hunting whales is now mostly a thing of the past. Of course the Japanese still do a lot of whaling for research purposes. Some of the research projects they're currently engaged in are: 'Does Whale Taste Good?', 'How Much Whale Can I Get On My Chopsticks?' and 'Have You Tried Some of This Yummy Whale Burger?'

Spirella corsets were displayed, fitted and sold in the client's own home. The fitting part was the fitting of the woman into the corset and not the other way round. It was not a delicate operation, not one that a lady could possibly undergo in any kind of changing room in a department store even if they'd existed then, which they didn't. Instead, my grandmother arrived much like a midwife, and the poor customer was then coached through a series of breathtakingly painful contractions until delivered into the firm embrace of the Spirella corset. My grandmother's father was a blacksmith with a strong back and arms like pistons. My grandmother inherited these physical attributes which served her well in the equally physical task of getting her clients shod in the Spirella iron lung.

An equally impressive physical feat was the fact that my grandmother sold her Spirella corsets from the back of her bicycle. Her sales patch was the United Kingdom south of a line between Inverness and Oban. The far North of Scotland was jealously guarded by a cranky Scottish widow who supplied the local highland gentry with corsets for nefarious sexual purposes, I believe.* My grandmother did, nevertheless, cover London and the South-East and thought nothing of pedalling a hundred miles on her bicycle for a couple of hard-fought corset fittings during the course of a single day.

Going to Loughton from Oxford to visit my grandmother was an exciting business. The M25 hadn't been built so it was a pretty easy run across country. It often meant travelling in my grandmother's car, a white Triumph Herald (my grandfather had a conscientious objection to driving). The Triumph sounded like a Spitfire and gave the impression of making a lot more progress than it actually was in pure distance terms. One of the reasons we didn't make as much progress as we might have done was the fact that, as a bad traveller, I felt the need to throw up approximately every three hundred yards. Even now, when travelling along those same roads, I often speed past a piece of verge which seems super-familiar because I was bent double over it for half an hour in my youth.

The reward for getting to my grandmother's house was a bowl of Rice Krispies. To me Rice Krispies with hot milk and sugar are the antidote to almost all of life's unpleasantnesses. Given their obvious therapeutic benefits, what baffles me now is the fact that we never, ever, had them at home. The El Salvadorian porridge regime

* Sorry. Total rubbish.

continued unabated until secondary school, where I was suddenly put on to two poached eggs on toast. Once rickets had been averted with the porridge, I suspect the eggs were introduced to prevent the early onset of Alzheimer's.

As part of our preventive medical programme at home we also had to have one spoonful of cod liver oil each night to keep every-thing moving. We were methodically lubricated in the same way that my grandfather would oil his lathe and for the same reasons. Following the cod liver oil, and by way of a treat, we were allowed to have one teaspoonful of rose hip syrup. Heaven only knows what this was for. Bearing in mind my mother's predilection for ancient herbal reme-dies, it would have been for some specific ailment I had been spotted to have or suspected of developing, such as short sinews or defe-cating follicles. With the cod liver oil and the rose hip syrup I must have smelt like a lightly perfumed trawler for much of my youth.

About thirty years later, after a spot of marathon running, my knees began to get a bit dodgy. The doctor helpfully suggested I take a regular dose of cod liver oil to keep the joints mobile. Personally, I think the problem was that the joints were overoiled and tended to slip all over the place. What the rose hip syrup has done over the long term I dread to think, although I do have an odd fascination for pagan rituals.

Rice Krispies, though, were a major source of excitement. I think that Rice Krispies were as exciting for my grandmother as they were for me due to the fact that they were just about the only packaged food she had in the house. I can imagine how modern and daring she would have felt requesting the newfangled Rice Krispies at the grocer's while at the same time weeping inside at the cost. Most of the contents of her larder was ripped directly from nature or bartered from local shopkeepers or poachers who themselves had ripped it directly from nature. It was unusual to arrive at my grandmother's and not be welcomed by a rabbit; not a little floppy number lolloping around the garden but a big dead one on the draining board. Most of her neighbours in Loughton had grown up in the surrounding country villages and it was still considered rude to visit someone and not bring with you a dead rabbit, a couple of birds, or, if you really wanted to make an impression, half a pig. When we arrived, our grandmother would catch up on all our news while she pulled the skin off a rabbit with the natural result that our news was conveyed extremely quickly in the barest possible headlines.

Aside from the dead rabbits, the other local wildlife found in Loughton was alligators. My grandfather had created this fiction that there were alligators under his shed and the only way to get rid of them was to put salt on their tails. (It was fortunate that Langton Machoko never visited my grandparents or he would have thrown a goat into the shed, if, that is, he'd managed to get it past my grandmother who would have had the skin off it before he'd got his own coat off.) My grandfather had lots of interesting things in his shed that he wanted to share with us, such as his beautifully oiled lathe.

In my view he'd created a bit of a rod for his own back with this alligator thing because I was loath to go anywhere near the shed and I remember thinking that my grandfather was being a little naïve if he thought that a pinch of salt was going to deal with the problem. The Fatted Calf didn't have such an active imagination so he spent much of the holiday with my grandfather in the shed banging nails into bits of wood and claiming the intellectual property rights on the subsequent inventions. I spent the time trapped between the rabbits and the alligators, reading Alastair MacLean novels.

When my grandmother wasn't skinning rabbits she was skinning local property developers. She had always had a very good head for business whereas my grandfather seemed to have a conscientious objection to money. I still remember the terrible, terrible incident when my grandfather decided to pay for the extended family to have a Chinese meal on the occasion of his eightieth birthday. He was a very generous man in every respect but he hadn't been allowed to handle money since the Depression (that's the worldwide economic downturn in the 1930s, not a famous patch of glumness on his part). The bill for the Chinese meal probably came to over £100 which, the last time he'd paid attention to it, would have been approximately sixteen times his annual salary. He went up to the desk to pay the bill and then disappeared. We thought he'd gone to the bathroom but later we found him wandering dazed and confused in the car park. What should have been a great evening ended with him in a shocked silence staring into the middle distance, convinced that he'd just blown his house, his pension and our inheritance on crispy fried duck.

I don't know why my grandfather wasn't more interested in money because he was very good at numbers and a clerk by profession. Clerks don't exist any more largely because computers have made them redundant. Clerks could add up long lines of figures in their head, make entries in ledgers with perfect copperplate writing and file and retrieve things with the aid of nothing more complex than a quill and lightly moistened finger. For the first half of his life my grandfather stuck doggedly to his job as a clerk in a local firm called Diggins. This firm combined the unusual occupations of general building and undertaking and, if you'd had to choose a name deliberately to reflect a cross between the two, you'd be hard pressed to come up with a more appropriate one than Diggins. There were very few families in Loughton over which, at some stage, my grandfather hadn't put a roof or six feet of clay.

But like Clark Kent, this Essex clerk had another, more impressive secret life. In the thirties he'd help set up the Peace Pledge Union which was the precursor of CND; he'd been heavily involved in famine relief in post-war Europe where the babies with swollen stomachs and bulging eyes in the heart-wrenching pictures had blonde hair; and he'd been one of the first members of a shadowy religious sect called the Brotherhood who were hippies in attitude, stockbrokers in dress and thirty years too early for the dawn of the age of Aquarius. Like many hippies he had to rely on a private income and his private income was his wife cycling around the lanes of southern England selling armoured underwear.

Then, in the late fifties, my grandfather had a life-changing experience: death. Well, not quite death but he did contract tuberculosis which in those days would kill you as soon as look at you. There was no real cure for TB apart from rest and fresh air.* Normally fresh air means exercise so getting it while resting was a pretty tricky one to pull off. The newly minted National Health Service came up trumps and instituted open-air wards for TB sufferers. This meant MRSA wasn't a problem but snowdrifts were.

Resting completely for six months, my grandfather took up painting. His first subjects were the landscape he could see from his bed, the view of which was unobstructed by any kind of wall. He proved to have a gift for painting which he developed in the art classes he attended on his release from hospital, although by the sound of it there was precious little physically keeping him in hospital apart from being tucked up very tightly. Sadly, the tutor in his art lessons died of a stroke (not brush-related) but this left a gap which my grandfather filled, and continued to fill, until he was ninety, when he too died with brush in hand.

The family fortunes were in the hands of my grandmother, which was just as well. She prudently invested the returns from her Spirella empire in property in Essex (she was the top Spirella saleswoman

* Not the same kind of fresh air that my father had on a regular basis, which wasn't necessarily good for your health.

in the South-East for ten years and was single-handedly responsible for the simultaneous precipitous decline in the whale population in that time). As all the little villages of south Essex were being devoured by the rapid expansion of London, it was a pretty shrewd investment. She managed to house most of her twelve brothers and sisters as well as herself. Not bad for a woman who went to school on stilts. If you want a good tip, she is currently looking at some very interesting property opportunities in the Yemen, which she sees as the next Dubai.

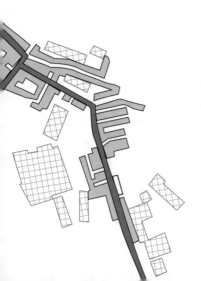

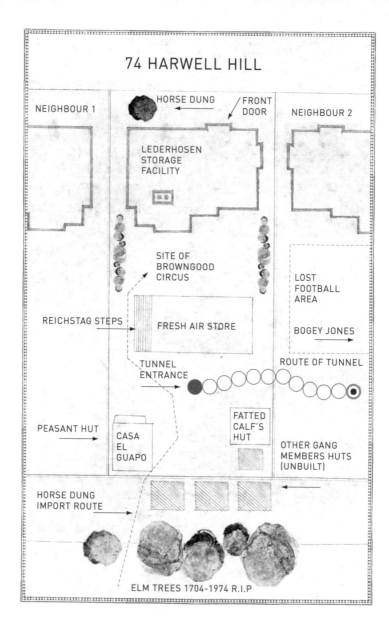

74 HARWELL HILL

NEIGHBOUR 1

HORSE DUNG

FRONT DOOR

NEIGHBOUR 2

LEDERHOSEN
STORAGE
FACILITY

SITE OF
BROWNGOOD
CIRCUS

LOST
FOOTBALL
AREA

REICHSTAG STEPS

FRESH AIR STORE

BOGEY JONES

TUNNEL
ENTRANCE

ROUTE OF TUNNEL

PEASANT HUT

CASA
EL
GUAPO

FATTED
CALF'S
HUT

OTHER GANG
MEMBERS HUTS
(UNBUILT)

HORSE DUNG
IMPORT ROUTE

ELM TREES 1704-1974 R.I.P

CHAPTER 9

Our Garden, Botley

One of the reasons my father decided on the house in Botley was that it had a very long garden. To put it in context, if Runny-Nose Barry had parked at the end of our garden we wouldn't have been able to see much more than the sun glinting off his slimy upper lip. My father wanted a big garden to grow his vegetables, more specifically the potato. After years studying Central American peasantry, my father had developed an almost mystical relationship with the potato. His idea of the perfect weekend leisure activity was to dig potatoes grown by his own fair hand. In some primal way it was his gesture of solidarity with the faceless peasant millions across the world who till the soil (he also really liked roast potatoes).

Many a time he tried to involve the Fatted Calf and me in the garden but my brother had an instinctive sense that wasting time working the land was how peasants stayed peasants. At his own primal level the Fatted Calf's unwillingness to get involved with the soil was a declaration of his capitalist belief that agriculture should be mechanised in order to release more value. I resisted the call to manual labour through sheer primal idleness.

We had a park at the bottom of our garden called Raleigh Park. It wasn't a swings-and-roundabout kind of park but a big hill studded with huge elm trees and horses and streams and mud – in other words the ideal children's playground. One day, while walking through this park (this would have been at the end of a Long March as walking round the park wouldn't have constituted much more than a stretching exercise for us), my father suggested that we boys might like to gather some of the copious horse droppings and bring them back to the garden as fertiliser for the potatoes. You can imagine how excited we were by this prospect but then my father added that we might be able to earn pocket money by doing so. My brother was staring at a fresh deposit of horse manure as this was announced and the expression on his face changed almost immediately in some kind of spiritual epiphany; what the horseflies were now landing delicately upon had suddenly transmogrified into a pile of the purest gold.

Half an hour later me and the Fatted Calf returned to the park armed with our seaside spades and a little hand-drawn truck to which the Fatted Calf had cleverly managed to harness me. After some very stiff negotiation with our father, we had agreed that for every truckful of horse manure a sum of one penny would be paid. That doesn't sound like much principally because it wasn't very much. But I think it may well have been an old penny which was worth sixteen Black Jacks and Fruit Salads, two kinds of sweets that were the principal cause of teeth being filled in my generation and then the fillings being pulled straight back out.

More important than the amount was the fact that real money was on offer. Our father had instituted another system of pocket money about a year before but at about the same time had very cleverly instituted a system of docking pocket money as a punishment for assorted misdemeanours. We weren't the best-behaved boys and our pocket money was docked at double the rate it was paid out so that, by the time of the horse-manure project, we probably owed him about £30 each in pocket money. The Sainted One wasn't given pocket money until long after because of the well-founded fear of how much she could earn in performance bonuses.

One truckload of horse manure equalled one penny; my brother had stumbled on the miracle of piecework and we hit the park with a vengeance. We filled six trucks without much difficulty and were mentally well on our way to becoming millionaires when we started running low on raw material. The whole park had been brushed clean and we were now beginning to follow horses round in the hope that the mood would take them. Then the Fatted Calf had another bright idea. Every so often the horses were taken out and ridden by the fat-bottomed women of the neighbourhood and all we would have to do was to follow the route they took and this would lead directly to the crock of gold at the end of the rainbow.

That's how the police found us in the middle of the A34 shovelling horse manure into a small cart. It was a busy road so the horses understandably got quite nervous before crossing it, which meant bonanza time for us. There was also a danger element to the collection process which the policeman pointed out to us after having coned off half the dual carriageway. But as my brother had shouted across to me from the hard shoulder, you have to speculate to accumulate. I expect he also explained all this to my father when the policeman took us home. My father was very upset that we'd brought the police home and he docked our pocket money severely, to the equivalent level of forty trucks' worth of manure.[*]

Although there were many rows of potatoes in our garden, there was also room for us children to do our own digging. When I was eight and my brother was nine we made a secret contract with Bogey Jones, a friend of ours who lived some way down our hill (no

---

[*] Something dreadful happened to Raleigh Park shortly afterwards, possibly occasioned by the removal of every scrap of horse dung. All the magnificent elms died. Every one. And suddenly the whole park was nothing but a marshy shrubbery. Trees have grown back now, although sadly not the elms. The horse manure has returned, too, and I still have to fight the compulsion to scoop it up and take it home with me.

relation to Runny-Nose Barry). We agreed we would both start digging in our respective gardens and then tunnel under the twenty intervening gardens and meet in the middle. Looking back on it, I am impressed by the fact that we never once asked ourselves why we wanted to do this. We solemnly undertook this massive feat of juvenile engineering for no good reason I could see other than having to avoid the tiresome stroll down the hill. I had little time to ponder my doubts for I was immediately delegated by the Fatted Calf to work at the face of the tunnel. After working all day I had sunk a shaft (using the correct technical jargon was as important as the digging itself) to a depth of six feet. As this was double my height I regarded this as a considerable achievement.

We decided that it was time for the horizontal shaft (gallery) to be started and so I rested on my plastic spade for a while as the Fatted Calf consulted our technical manual, a copy of *Escape from Colditz*. I was ordered to continue digging while he ripped boards from my bed to shore up the tunnel. The boards from *his* bed would have been too big and not the correct type of wood. My enthusiasm for the tunnel waned as teatime neared. At precisely ten minutes to five I was relieved at the tunnel face by the Fatted Calf, who proceeded to dig furiously and get himself amazingly dirty in five minutes. His timing was perfect. My father strolled down the garden and called us both in for tea, whereupon the Fatted Calf emerged triumphant from the tunnel, mud smeared judiciously on his forehead and blinking as if unaccustomed to the light. I was standing at the top looking as if I'd done precisely nothing all day.

Over tea my mother asked why we were digging. Had we lost something? We weren't planning to bury our little sister, were we? My brother patiently explained that we were digging a tunnel down the hill. When asked, equally patiently, by my mother why a tunnel was necessary, he replied with great earnestness: 'Because there isn't one there', which as far as explanations go is up there with climbing Everest 'Because it's there'. After tea the Sainted One contrived to fall into the tunnel on purpose and we had to fill it in (after extracting her). There's still an indentation where the tunnel was dug because there never seems to be as much mud to put back in holes as you take out. I don't quite know how that works but it would be a hell of a problem if we're forced to fill in the Channel Tunnel.

I've always loved tunnels; I'm sure Sigmund Freud would have had something to say about that. Like Freud, I had a one-track mind but in my case the track was OO gauge and the only thing on it was the *Flying Scotsman*. There is something profoundly satisfying about railways that jink through mountains by a series of tunnels and bridges. Superhuman efforts of engineering are needed to persuade the mountains to yield a gentle gradient along which a train may travel and I tried to incorporate this into my own model railway.

I built a papier-mâché mountain landscape of vertical inclines, plunging gorges and impassable ravines. Through all this my *Flying Scotsman* ran, lost to sight most of the time in endless tunnels and popping out occasionally to speed through a model German village. It never worried me that there were no stations or people in my layout, nor did I worry that the *Flying Scotsman* was chugging through a Himalayan landscape dotted with Germanic villages. But tunnels and bridges do that to you; they remove you from reality just as, after we filled in our tunnel in the garden, we forgot to tell Bogey Jones, who continued digging for five days.

The Bogey Jones tunnel wasn't the only hole in our domain. Halfway down the garden was a big crater filled with rubble. Something made of red brick had been knocked down somewhere nearby. I'm not sure what this thing was as our house seemed structurally intact. Maybe it was an ambitious bit of fly-tipping by a local builder, although, given the fact that we couldn't get a ton of horse manure round the side of the house, he would have had to make an awful lot of surreptitious trips with a wheelbarrow to fill the hole with rubble.

The top half of our garden was given over to flowers, which were my mother's department, while the bottom half was vegetables which came under the direct authority of my father. My father would no sooner suggest a flower for the top half of the garden than my mother would suggest a vegetable for the bottom. The closer my father got to the bottom of the garden, the more he became like one of his beloved Central American peasant farmers. This was where the potatoes and tomatoes and maize grew and he would wear a peasant's straw hat while toiling in the rows among them. He'd also built himself a little peasant hut where he could go and sneakily test the tobacco crop. Between my mother and father they soon began to get the garden into shape and the pit full of bricks and rocks became increasingly incongruous in the no-man's-land between flowers and vegetables.

Eventually it was decided that the pit should be cleared. Money was attached to the project and the Fatted Calf's attention was immediately engaged. By an absolute stroke of luck I was given the job of removing the rubble, while the Fatted Calf thought about what to do with the hole. This hard creative work took considerably longer than he anticipated but it did give me the chance to finish removing all the rubble. Using various primitive hoists and barrows and my limited brute force, I transferred the rubble to the very end of the garden to form part of a wall against unsavoury elements entering, such as fly-tipping builders.

By the time the hole was clear the Fatted Calf announced his master plan which was to create a sunken garden. This would comprise steps leading down into the pit with a tiled seating area at the bottom.

The garden element would be a sloping rockery on each side of the pit. Of course you can't build a rockery without rocks but fortunately we had a whole pile of them at the other end of the garden.

Once I'd finished the 'simple logistical task' of retrieving the rocks from the far end of the garden,* the hard work could start. My brother had ordered in some concrete, breeze blocks and paving slabs. This was going to require some very hard grown-up labour indeed so my grandfather (the conscientious objector) was called in to help and didn't manage to object fast enough or conscientiously enough. After a few days of building the work was complete and my parents were summoned to see the result. If I remember correctly the reactions were mixed. Three sides of the pit were covered in rocks, concreted in place with more concrete than the rocks themselves. On the third side were steps that would have graced the Reichstag in Berlin. These formed a massive bomb-proof staircase where practicality had clearly been sacrificed to indestructibility as there were only three steps and each had a drop of about two feet.

In normal circumstances my mother would have taken one look and told my brother to clear up the mess and then go to his room. My mother yields to few things but concrete is one of them. It was generally agreed that the pit was a vast improvement on the pile of rubble and that my brother might usefully employ his new-found skills elsewhere, the more elsewhere the better. The sunken garden is still called the pit but, all credit to my mother, plants have begun to cover the concrete cladding on the sides. It was also used a lot more than initially anticipated because my father found a source of fresh air at the bottom, out of the direct line of sight from windows at the back of the house.

---

* 'Simple logistical task' was the Fatted Calf's way of describing back-breaking physical toil when assigning it to me. Interestingly, when he had to undertake it himself, he reverted to the phrase 'back-breaking physical toil'.

One thing that slightly galled me about the pit was that the Fatted Calf's last job was to set his name in concrete at the foot of the steps in foot-high letters. By the time I had realised what he had done, and the fact that my name was nowhere to be seen, the concrete was dry. I found one thick bit of concrete on the rockery which was still damp and managed to carve my initials in letters about seven millimetres high.

If you've ever had a tour of Oxford you'll know that one of the finest buildings is the Radcliffe Camera, a circular library built by Sir Christopher Wren. Many years later I happened to find myself on the roof of this building which was totally out of bounds to everyone except the person I was with, who was one of those very useful people who has a key to every door in Oxford. Carved into the lead lining on the roof, out of sight of everyone except the once-in-a-century visitors like ourselves, were the names of all the men who'd laid the roof – men with good Oxfordshire names like Woodford, Belcher, Newing, Froud, Farmer, and Wright. In a slightly similar way, everyone knows that the Fatted Calf was the architect of the Reichstag Steps but a select few (i.e. me) know who did the hard labour.

Writing his name in concrete was the culmination of a habit the Fatted Calf had been fostering for some years. While developing as a foetus, he had acquired an instinctive grasp of property law. In the first few months of life outside the womb, he'd quickly understood that possession was nine-tenths of the law and, to make sure this worked to his advantage, he developed a grip that would make a vice weep. The first word the Fatted Calf uttered was 'mine' and the first words he wrote on paper were 'This Paper Belongs to F C Browning'. He then practised writing this phrase on everything that had a flat surface and was likely to be of value in an open market. Things came to a head when he discovered Humbrol enamel paints and proceeded to paint his name on all the tools in our father's shed, reasoning that if my father hadn't bothered to write his own name on them, then he obviously didn't want them.

The Reichstag Steps was one of many grand infrastructure projects in the garden instituted and managed by my brother. An earlier project had been a local mail service after the Fatted Calf spotted that the Royal Mail was a monopoly ripe for competition. We built an entire sorting office to serve every street on the estate (I still have the file for Laburnum Road behind me[*]). This service was developed in conjunction with our friend Mark Good who lived in Laburnum Road. We didn't have any mail to sort and by the time tea was called, our interest in the job waned. Mark was a great business partner and we had another idea to run a circus from our back garden. This was to be called the BrownGood Circus. We probably considered the GoodBrown Circus but my brother would no doubt have had some pretty strong views on branding. The sign was made and painted but I don't think the wild animals ever arrived, at least not before tea.

The most grandiose project was to build living accommodation for all six people in our gang right at the bottom of our garden. Each of us was to have a separate detached hut in its own grounds. This idea must have been inspired by our holiday in the Kidwelly Refugee Camp and, had we finished construction, living conditions would have been every bit as good and the surrounding environment many times better. First, of course, we had to build the show home which also happened quite by chance to be the one reserved for occupation by the Fatted Calf. After his house was built, the slave labour all started to get a bit cross and I led a Spartacus-style revolt. A stone was thrown, the Fatted Calf was hit and blood flowed from

[*] It has 'This File Belongs to F C Browning' written in Humbrol paint on it so I'm assuming I only have it on a very long lease.

the blood-filled heads we Brownings seem to specialise in. The fallout from that incident was so appalling that I've completely forgotten it, although I expect the memory could be retrieved by experts. But my memory is one of the few things I'm not keen to tunnel too deeply into.

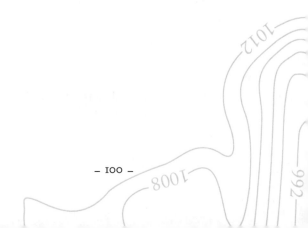

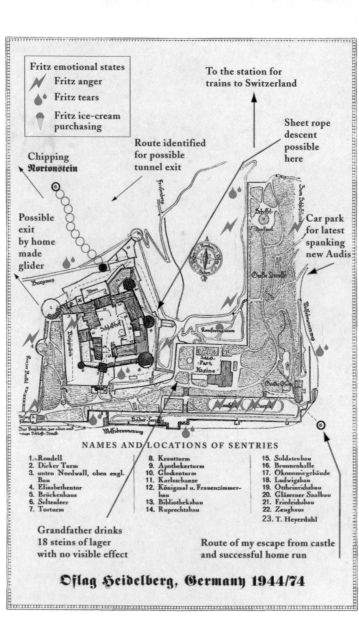

Fritz emotional states

Fritz anger

Fritz tears

Fritz ice-cream purchasing

Route identified for possible tunnel exit

To the station for trains to Switzerland

Sheet rope descent possible here

Chipping Nortonstein

Possible exit by home made glider

Car park for latest spanking new Audis

NAMES AND LOCATIONS OF SENTRIES

1. Rondell
2. Dicker Turm
3. unten Nordwall, oben engl. Bau
4. Elisabethentor
5. Brückenhaus
6. Seltenleer
7. Torturm
8. Krautturm
9. Apothekerturm
10. Glockenturm
11. Karlsschanze
12. Königsaal u. Frauenzimmer-bau
13. Bibliotheksbau
14. Ruprechtsbau
15. Soldatenbau
16. Brunnenhalle
17. Ökonomiegebäude
18. Ludwigsbau
19. Ottheinrichsbau
20. Gläserner Saalbau
21. Friedrichsbau
22. Zeughaus
23. T. Heyerdahl

Grandfather drinks 18 steins of lager with no visible effect

Route of my escape from castle and successful home run

Oflag Heidelberg, Germany 1944/74

CHAPTER 10

Oberheinriet, Germany

I grew up believing that this country's finest hour was the defeat of Germany in the Battle of Britain and the happiest days of my childhood were spent recreating this and other battles in the playground at Elms Road Primary School. I was encouraged by my fellow combatants in the playground to believe that the only good German was a dead German, but I found this hard to go along with because we knew Fritz, who was genuinely German, had genuinely fought in the war but, apart from those two major black marks against his name, didn't appear to be that bad.

Nevertheless, I spent most of my childhood totally immersed in the Second World War. In primary school we were once asked to do some creative writing on travel. I chose to describe Operation Pedestal, the vital wartime convoy through the Mediterranean that prevented Malta succumbing to Germany and threatening our entire operations in North Africa. There was a map, of course, describing the exact route of the convoy, and if it hadn't been put forward for a prize to the National Historical Association it would have been republished here. I'm afraid I'm lying about the National Historical Association but not as much as you'd imagine. My father also spent his childhood obsessed with the Second World War partly because it was being fought over his head in London. Occasionally, when some aspect of my homework interested him, he would bring the full resources of Oxford University to bear on it, which probably explains why some of the mapping on my primary school creative writing project is still being used to explain that theatre of war to undergraduates. In any case I certainly got a gold star for it.

By the time I was ten I was in the full flood of my obsession with the Second World War; I could read adult books on the war and had plenty of time to do so as I was not yet interested in girls and other peacetime activities. It was precisely this moment when my parents chose to send me on a mission to Germany which they lightly classified as a 'holiday'. Of course this sudden long-range mission into the heart of industrial Germany didn't just appear from nowhere. It came from a negotiation involving the Fatted Calf. Earlier that summer my mother had taken my brother on the trip of a lifetime to visit friends in Boston, Massachusetts. My mother's best friend's eldest boy was best friends with the Fatted Calf so it worked well. The best friend's second son thought I was an idiot, so I wasn't invited. As he knew very little about the Second World War we would have had very little to talk about, so it's probably just as well.

One of the cast-iron rules my father lives by (unless there are extenuating circumstances) is absolute parity of treatment between the children. This was underpinned by a heavy foundation of socialist philosophy that each child should contribute according to his ability and receive according to his needs. He must have calculated that exposing my brother to the unfettered rawness of capitalism in America would cure him of that inclination and that sending me off to Germany would likewise give me some kind of closure on the Second World War. Sadly, socialist theory rarely works in practice and the trip to America put the Fatted Calf firmly on the road to Wall Street and capitalist excess. My trip to Germany did succeed in wrenching my focus from the Battle of Britain but only as far as Colditz, other prisoner-of-war camps and the subject of how to escape from Germany and get home, which may give you an inkling of what kind of holiday that proved to be.

I travelled to Germany with my grandfather, the conscientious objector.* We were picked up from Stuttgart station by Fritz in his brand-new Audi. This was a little bit disconcerting. If we had won the war how come my father was driving a hand-welded cut-and-shut death trap and Fritz, who, after all, had lost the war and had been our prisoner for some time, was driving a brand spanking new Audi 80. Of course there were complex social and economic reasons for this but I have never been one for social and economic complexity and certainly wasn't when I was ten. It turned out that Fritz worked in the Audi factory and, like a chocolate factory, they let you take home all the Audis you wanted until you were sick of them (I'm not sure that's strictly true). But it was noticeable on the trip back to Fritz's village that there were a lot of Audis on the road but none of my father's VW Vantages, even though my father had often assured me that our car was the latest technological marvel fresh from Germany, the front half at least.

Fritz lived in a small village called Oberheinriet, which I was absolutely certain was German for corporal. This suspicion was strengthened by the fact that the neighbouring village was called Unterheinriet, which surely meant lance corporal. The region was called Untergruppenbach, which means staff sergeant in the province of Baden-Württemberg, which I know means self-propelled artillery piece. All the names I came across were instantly familiar from my weekly subscription to *Commando* magazine but I also found that the villages themselves were strangely familiar. This was because the houses were exactly the same as the ones on my railway layout at home.

The nerdy trainspotters among you, that is to say the men, will know that most model railway equipment comes from Germany

* My parents had decided not to come on this mission: combat fatigue probably.

(apart from the Hornby *Flying Scotsman*, which comes from China). If you want to build a village for your train to roar through, rattling the windows and covering the laundry with soot, you have two options. Either you can take ten years to lovingly recreate a perfect scale replica of an English cottage using spit, matches and ear wax, or you can buy little houses from the model shop, all of which will be perfect replicas of modern German houses in Unterheinriet, Oberheinriet or Untergruppenbach. I had several of these houses on my railway layout and it always seemed strange that the *Flying Scotsman* steamed through a picturesque German village. It created a picture of post-war Europe that I simply couldn't get my head round and was one of the reasons why my entire layout ended up speeding through long underground tunnels out of sight of any kind of habitation.

Oberheinriet turned out to be a very pleasant village lying on a steep, vine-covered slope. Driving slowly up through the village I was immediately struck by two things: firstly, and most exciting, was that many of the houses seemed to have cows and other livestock in their living rooms or at least where their living rooms would have been if the houses had been built in a sensible English fashion. My family and our Spanish guests in Botley were used to tons of steaming manure outside the front door but inviting the actual producers of the manure into your living room was a touch of eccentricity that none of my reading of the German character thus far had prepared me for. Secondly, at roughly the same time I noticed that moving around the village were a number of old women, all dressed in black, pulling handcarts. This was a hugely exciting development as I assumed they were carting dung around the village, possibly up to the vineyards for a few pfennigs. Later it turned out that these ladies were on their way to and from the communal oven where all the villagers baked their bread. This struck me as superbly primitive and more in keeping with defeat in the Battle of Britain.

When we arrived at Fritz's house I was slightly disappointed that there was no livestock to greet me downstairs in the traditional Oberheinriet way. Fritz had built his own house in a more modern style employing some of the heavy construction skills he'd perfected making his living-room furniture in the POW camp. Instead of livestock in the front room there was a young German boy my own age called Ralf. He was the Fatted Shnitzel of Fritz's family. After the initial introduction we were left to play by ourselves for the two weeks left to us. This immediately threw up some very serious complications. I had four favourite games at that time: Battle of Britain, Normandy Invasion, Desert Army and Bomber Offensive on the Ruhr. In all of these games there was, of course, a ready-made role for Ralf but I just couldn't bring myself to introduce him to any of them as I sensed it would be abusing his hospitality. I can't remember what we did end up playing. It probably wasn't a 'supra-national economic community across Europe' kind of game as that would have had limited play value. I may have borrowed the bread trolley and introduced him to dung collecting, which would have changed the flavour of the bread in the village for while.

Fritz was a popular man in Oberheinriet (at least until the bread poisoning incident) and everywhere we went we were treated like royalty. Fritz himself had three basic modes of being: extreme anger, floods of tears or ice-cream purchasing. Often you would get the first two during the last one as he seemed to find buying an ice cream a very emotional experience. When Fritz was working in the fields after the war he would visit my grandparents' house for lunch on Sunday and never fail to bring with him a block of Neapolitan ice cream which, in time of rationing, was an extremely generous and impressive thing to do, especially for my mother and her sister who looked on him as an older brother. For obvious reasons Fritz had got it into his head that the way to manage the English was to give them ice cream, in much the same way José had learned that it was best to have a bath and carry a tapioca pudding to make any kind of impression with the locals. In the two weeks I was there I was shown all the local sites which I am told included Rotenberg, the Black Forest and Heidelberg. I can recall very little about any of these lovely places principally because I couldn't see past the colossal ice cream that was in front of my face for much of the time.

I remember my homecoming from Germany more clearly than I remember most of the time away possibly because I didn't go straight home but instead went to meet my parents at the home of an elderly Swiss gentleman, a long-time friend of the family. I have a bronze bust of the Swiss Gentleman in my office now and by the look of it he was a very handsome man in his prime. He was also an extremely wealthy man as his family had invented some kind of electrical thing, possibly a transformer, which was vital to just about everything that needed electricity. His house was extremely large and everything in it was of the absolute highest quality.

I remember this because my parents had given us children strict instructions not to touch anything anywhere in the house. The sofas especially were a no-go area; they were vast constructions that had cushions stuffed with the finest Hungarian goose down. This made them extremely comfortable but also immensely squishy. A moderately vigorous fart would leave an indentation as if you'd dropped a medicine ball on to the sofa from a great height. We had also been told that the Swiss Gentleman didn't like children. This put him very high in our estimation because my brother and I knew instinctively that we weren't particularly quiet, clean or likeable. The Sainted One was a different matter but I don't think an exception was made even for her.

Given the fact that the Swiss Gentleman was a man of impeccable taste and breeding, we could never really explain the presence and atrocious manners of his mynah bird. At one end of the cavernous, wood-panelled entrance hall of his house there was a large bird cage with a fresh copy of the *Zürichsee-Zeitung* laid on the floor and a single, mangy looking mynah bird perched above it. Mynahs sound exotic but they're really only starlings with a passport. They're not particularly pleasant to look at so maybe the attraction is that, like children, they can be trained to speak. The Swiss Gentleman had obviously spent some quality time with the mynah bird because he'd trained it to say 'Guten Tag!' possibly to alleviate feelings of homesickness for his native Switzerland. Unfortunately the programming had gone slightly awry because, although the bird could say 'Guten Tag!' with gusto, he only said it when he was simultaneously defecating on his newspaper. I know of nowhere in the world where that is considered good manners.

There was one other unexplained feature of the Swiss Gentleman's domestic set-up that has troubled me for far longer than the tatty little mynah bird. Often people leave in attics some little item that gives you an insight into their personal life, and so it was with the Swiss Gentleman. Firstly, you have to imagine that his attic was the size of our house in Kidlington as we boys could run screaming from one end to the other in the time it took my father to rush up three flights of stairs and tell us to belt up before the Swiss Gentleman's studied air of neutrality slipped. Right in the middle of this long, carpeted attic was half a billiard table. This billiard table was obviously one of the finest ever made. For a start it was full size, which made it half the size of a squash court (or quarter of the size because there was only half a table). The legs were beautifully carved wooden lion's paws and everything reeked of quality and expense. You can imagine how excited we were to find the table until we found that exactly one half of it had gone missing.

We knew even at that age never to ask direct questions about things in attics, so, after some very oblique enquiries, we learned that the other half of the majestic billiard table was in the Swiss Gentleman's workshop where he used it as a workbench. As cutting the table in half would have been an epic job in the first place, this remains a complete mystery to me. Maybe it was sold to him by the same man who sold my father the car of two halves.

Apart from the incontinent mynah and the half-billiard table, the Swiss Gentleman was clearly a bit of a dude. During the First World War he had been an officer in the Swiss Balloon Corps. Obviously the Swiss didn't fight in this war, or any other since 1815, which is why they think a multi-purpose knife and a balloon corps is an effective defence strategy. Nevertheless, being in the Balloon Corps was clearly a pretty dangerous occupation because in mountainous Switzerland you have to go a long way up before you can move sideways and then, if you're not careful, the merest gust of wind can land you in a hot spot like the French Riviera. My grandmother had a photo of the Swiss Gentleman taken during the First World War (not by her, as she was still on stilts in the Essex mud) where he is in full Balloon Corps uniform sitting in a nifty little goat cart. I have talked about carts and trolleys before and I should stress that this cart was for a different and more hygienic purpose. It was a one-man cart pulled by a goat. In those pre-motor car days, the goat cart was for the officers of the Swiss Balloon Corps what the MG sports car was for the later Battle of Britain pilots. On their goat carts they could get from their quarters to their airfield in a matter of minutes if the balloon went up (as it were).

After a few days with the Swiss Gentleman, we finally returned to Botley. That trip to Germany has stayed with me. If I'm *totally* truthful, a little part of that trip to Germany has stayed with me, or at least quite close to me. I was given a parting present by Fritz and his family as a reminder of my stay in Germany. I am happy to tell you that it was not a bemused looking doll in Bavarian lederhosen. I am less happy to tell you that it was a real pair of lederhosen. This was long, long before leather trousers became an acceptable form of work clothing in the media and, even now, leather shorts would probably raise a few eyebrows. At home, in the Fatted Calf's bedroom there was a disused fireplace which my mother had made into a bit of a feature by filling the grate with oversized pine cones. I made it an extra special feature by ramming the lederhosen as far as I could reach up the chimney. There are pine woods behind Fritz's house in Oberheinriet and for years afterwards the pine cones in the grate reminded me that there was some quality German leatherwear wedged not far above them. The lederhosen are still up the chimney and one day I hope some archaeologist will dig them out and come up with some ludicrous theory based on his find.* Looking back, I admit that stuffing them up the chimney was a selfish and ungrateful thing to do but then war does bring out the worst in people.

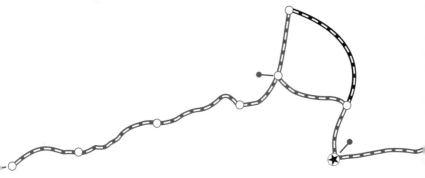

* Archaeologists base most of their reading of prehistory on things they've found and dug up. Bearing in mind that most things that have survived for a long time have been deliberately hidden for odd reasons, it seems a very shaky basis for writing history. One day there will probably be a theory that ten-year-old German boys swept chimneys in Botley in the late twentieth century.

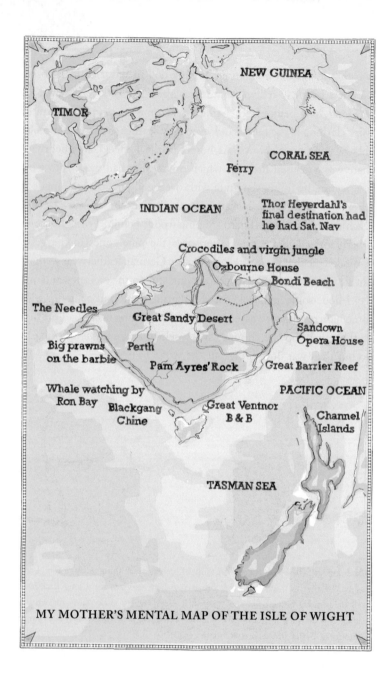

MY MOTHER'S MENTAL MAP OF THE ISLE OF WIGHT

CHAPTER II

Isle of Wight, the English Channel

In the winter after my mission to Germany I fell ill with German measles. To me there was a clear and obvious correlation but apparently at a microbiological level there isn't. In close succession I also managed to contract scarlet fever and glandular fever. As you can imagine I was red and hot and Germanic for some time. The doctor ordered me to have complete rest but, thank goodness, no fresh air. This meant I was in bed for about two months which sounds great when you're an adult but which is fantastically dull when you're a child.

My mother had simultaneously gone into a period of shock, trying to come to terms with the failure of her porridge/cod liver oil/rose hip syrup health regime. However, she and my father then realised that they had been presented with a golden opportunity: I was approaching the end of my primary school years and the 11+ exam was looming. My parents couldn't afford to send me to private school but it used to be the case that, if you passed the 11+, the Government would very kindly pay for you in the belief that you would eventually return the favour in taxes. As the highest rate of tax was then 97 per cent that was probably a reasonable calculation.

Before long I had so many books on my bed that there was a real danger of me adding collapsed lung and cracked ribs to my other ailments. The only books denied me were anything to do with the Second World War as by that stage I could have written my university dissertation on practically any aspect of the Second World War

(I did later). It took me a long time to recover from my various illnesses but by the time I was fit, I also had a brain the size of Jupiter. Given this huge advantage I sailed through my 11+ and got a scholarship, much to the relief of my parents. In the last few weeks of my recovery, my mother decided that fresh air was finally in order and organised a convalescent holiday to the warmer climes of the Isle of Wight.

The Isle of Wight is basically an economy Australia without the tedious long-haul flight; at least, that's how it was sold to me. Because you take a ferry across to the island you subconsciously feel that things are going to be different on the other side. They were indeed very subtly different in that there's nothing on the island that doesn't also fit on a ferry. When we arrived there on a freezing day in May, I realised that the Isle of Wight was actually a lot more exciting than Australia because it had underground trains as overground trains. A little London Tube train started right at the end of the pier (even more bizarre and exciting – a tube train at sea!) and then wound its way through Ryde, Shanklin, Ventnor and other oddly named places. This overground underground is such a good idea I don't know why they don't try it in other parts of the country. Underground trains are small and cute and hold lots of people, and if you run them overground you don't even have to bother with a tunnel. I say put them in bus lanes and get rid of buses, which are far less exciting.

In Ventnor we stayed at a bed and breakfast. That may not sound very exciting to you, but to me it was like rocking up at the Burj al-Arab. Beds I was more or less familiar with but breakfast was a holiday in itself. Porridge wasn't on offer (that alone enormously accelerated the convalescence) but Rice Krispies were! And you could just help yourself without having to make the trip to your grandmother's, throwing up five times on the way. Incredibly that was only the starter. You then had a choice of eggs, bacon, tomatoes, baked beans, potatoes, sausages, mushrooms and fried bread in something known as the Full English. If that's what it means to be fully English then thank the Lord I was born in this wonderful country. After that lot had been finished, there was dessert of toast accompanied by marmalade in little pots. Those little pots are such fun I don't know why people don't have them at home all the time instead of those big ugly jars. I think that breakfast was also a bit of a turning point in my mother's life because, when we returned home, poached eggs usurped porridge at the breakfast table.

To be honest I felt slightly the worse for wear after finishing my first Full English breakfast, but fresh air soon steadied me. As my father wasn't with us on the Isle of Wight, Long Marches were off the menu. Besides, in my weakened state I wouldn't have been able to manage more than the first ten or fifteen miles. Instead we got our fresh air on the top of the island's many open-topped double-decker buses. They should have put TB patients on the top of the 64 bus between Blackgang and Ryde; you got so much freezing cold air they would all have been discharged, completely healthy, by Ventnor.

One day rain added dangerous dampness to the fresh air and my mother decided that we should go and see a film. It was showing in Ryde and we were in Ventnor so we had to take the open-top 64 bus. We were soaking by the time we got there but chock full of fresh air. Why we couldn't have sat downstairs, I'll never know. The film we saw that day has always stayed with me: it was *Breakheart Pass*, starring Charles Bronson. It was all good healthy stuff with railways and fresh air (much like our holiday) but what we didn't realise was that there was a short film before the main picture. In my memory it was thirty minutes of hard core pornography. In reality it probably featured two people kissing politely although I do remember a bra strap on screen. I could have handled this kind of graphic imagery on my own but, being sat next to my mother, it was very hard to cope with her embarrassment. Apart from that setback, the holiday passed off peacefully and I returned to Botley fit and finally ready for further education and my first pair of long trousers.

My next visit to the Isle of Wight came five years later and was one of the first holidays I was allowed to go on without my parents. All credit to my parents, they have never been shy about me leaving home often for long periods. Indeed, they often look mildly surprised when I come back. In the intervening years I had discovered Youth Hostels, which are a fantastic institution allowing young people to sleep safely and warmly in lots of lovely parts of the country. You have to prepare your own meals and help with chores around the hostel but it's a great way of having a relatively comfortable adventure (and also learning what a chore is and how to prepare and cook your own Weetabix).

My first Youth Hostelling trip had been with the Fatted Calf. We decided to take the train from Oxford to Worcester and then cycle back, breaking the trip at four Youth Hostels in between. We cycled into Oxford station and caught the train, no problem. It's a lovely ride up to Worcester and I'm sure the Fatted Calf sitting up in First Class enjoyed it as much as I did. At Worcester we had our first cup of tea, paid for with our own money. Personally I didn't have any money because of the huge pocket money indebtedness situation but the Fatted Calf had plenty thanks to his many business schemes, and he very generously lent me some at an attractive rate of interest. From Worcester the plan was to cycle down to Malvern, where our first Youth Hostel was situated. Not having done this sort of thing before, we cycled the ten miles in about half an hour and then found ourselves at the door of the Youth Hostel some considerable time before lunch and about six hours before opening time.

This is where our problem with under-age drinking began. Having discovered there was nothing to do in Malvern apart from take the water and play Space Invaders on a machine in the station buffet, we resolved not to make the same mistake the following day. After a late start and about an hour's extremely leisurely cycling we stopped at a country pub and my brother managed to procure some cider. At this stage I need to tell you about my relationship with alcohol. To be blunt, it's dysfunctional simply because alcohol doesn't work on me. I could drink myself to death and still not feel in anyway light-headed or even slightly relaxed. This is something I inherit directly from my grandfather, the conscientious objector. He, too, was totally impervious to alcohol and would regularly drink hardened German war veterans under the table. I often think this did much to cement his heroic status with the good burghers of Oberheinriet. He would regularly join in any drinking games, confident in the knowledge that he would not only win but that it would have zero effect on him. He was the most hard-drinking model of sobriety I have ever known.

Sadly, this gene passed directly to me. I can drink anti-freeze from the can and totally fail to warm up socially. Neither of us knew this as we faced our first cider in the garden of a lovely pub high on Cleeve Hill in Worcestershire. I can't remember whether we had one cider or two but I do remember what happened next. Feeling rather manly and worldly-wise, we both got back on our bicycles and headed down the hill. Cleeve Hill is rather a steep hill and we free-wheeled down it with gay abandon, indulging in some un-English Texan-style whooping on the way. At the bottom of the hill the road bent gently round to the right but the Fatted Calf didn't. He continued straight ahead at full speed into the field at the bottom of the hill and I found him lying unhurt and giggling in the long grass. He couldn't be persuaded to move so I think we stayed there for some time, my brother asleep and me wondering what had brought on this sudden relaxed mood and unusual lack of target focus in the Fatted Calf. At least we weren't four hours early at the next Youth Hostel.

One thing that the cider did help us with was the relative speed of our bicycles. My brother had a beautiful Claude Butler Five-Speed Racing Machine. Nowadays your basic starter bicycle comes with thirty-seven gears but five of them was pretty louche when I was growing up. The Claude Butler was, not surprisingly, the best bike money could buy so my parents wouldn't have countenanced it were it not for the fact that it was largely financed by the profits from the Fatted Calf's Jammie Dodger distribution business. I, on the other hand, had some kind of sit-up-and-beg police bike with straight handlebars, a three speed Sturmey-Archer gearbox and a deeply naff leather satchel permanently attached to the back of the saddle. It pains me even to think about it.

This bike was called a Raleigh Wayfarer but it might as well have been called a Raleigh Wafer for all the street cred it had. It was humiliation on two wheels and what made it worse was that I paid £36 for it out of my Post Office Savings Account. This was real money given to me by relatives for Christmas and birthdays that had somehow been ring-fenced to avoid it disappearing into pocket money debt relief. I wanted a racing bicycle more than I can tell you, and at one stage I even tried to remodel the straight handlebars with a papier-mâché drop-handlebar attachment but that's too embarrassing to relate here. However, once the Fatted Calf was cider-powered, straight-line progress seemed to be difficult for him so I, sober as a judge, was able to keep pace with him for once on my horrid Raleigh Wafer.

As I was saying earlier, my second trip to the Isle of Wight was a Youth Hostelling holiday. I told my two best friends at school about Youth Hostelling and suggested that a walking tour of the Isle of Wight might be a terrific adventure. At first they were keen but then I think their parents said no. One of my best friends had the feeble excuse that he was going to Abu Dhabi for three weeks but the other told me quite candidly that his parents thought it would be too dangerous. This turned out to be remarkably prescient as we had two close shaves with death on that holiday.

As my two best friends pulled out, two other second-tier friends volunteered to go with me. I'm not sure how this happened because by that stage we were all going through our musical puberty. These two friends were very into heavy rock – Purple Sabbath, Black Floyd, Deep Pink etc – while I was at the same time discovering the deeper, complex and more lasting pleasures of disco. This fact alone made our friendship increasingly unlikely. One of these friends I will call the Ginger Rabbit because that's what he looked like. The other I will call the Bridge because he carried a guitar with him at all times, worshipped Jimi Hendrix and was forever saying 'Take it to the Bridge', which to me said instructions to train drivers but to him said legendary guitar playing. I'm quite keen not to reveal their identities because I happen to know that one of them is now a vicious corporate lawyer in the pay of Russian oligarchs. The other one I nearly killed, so if they were to get together now I might end up in prison.

The Bridge's father got into the swing of our holiday even before we started. He decided that if we were going to be Youth Hostelling in 'hostile territory' (I don't know what he had against the Isle of Wight) then we needed to be able to survive in the wild. To this end he blindfolded the three of us, made us lie down in the back of his estate car and then drove us to an unknown location. I suspect that if he did something similar today the police would make very heavy weather of it. From this unknown spot we had to find our way back to their house using nothing but a compass, an Ordnance Survey map and our basic survival skills. We took our blindfolds off, watched the Bridge's father drive off into the distance and realised we were about thirty yards from the Ginger Rabbit's house. Our basic survival skills kicked in and we went there for tea before catching a bus back to the Bridge's house. We all agreed we wouldn't tell the Bridge's father what had happened as we didn't want to ruin his fun, and not ruining his fun was also, the Bridge assured us, a basic survival skill in his house.

Having been suitably trained, we departed for the Isle of Wight. Our first Youth Hostel was near Southampton in a small village thrillingly called Botley, not as fresh and unique and wonderful as the original Botley in Oxford, but a Botley nonetheless. I was interested to see whether the locals had found a way of pronouncing it that made it sound slightly more refined. They hadn't. The following day we did the epic crossing to Ryde on the ferry and once again boarded the famous Isle of Wight Tube trains, me feeling like a regular commuter.

The second day on the island was going to be our own Long March. We would walk from Ventnor to Totnes, virtually the entire length of the island and a good twenty miles. This was peanuts for me but for the first time I did have a large pack on (I didn't need a large pack but for reasons known to me then I'd packed a primus stove, even though Youth Hostels are well equipped with cooking facilities). That morning we shouldered our packs and by way of a little joke the Ginger Rabbit stuck his thumb out. Even before his thumb was fully erect, a car had screeched to a halt directly in front of us and half an hour later we found ourselves sitting in Totnes wondering what to do for the next eight hours.

The next day improved in that we didn't get a lift so we could walk but got worse by virtue of the fact that a madman with a knife tried to stab us. The precise details are hazy but as this man walked past us the Ginger Rabbit made a comment and the next thing we knew we were running down the street being chased by a shouting, knife-wielding crazy man. It's a bad thing to say, but I don't entirely blame the man with the knife as there was something about the Ginger Rabbit that made you want to stab him. Shortly afterwards I, too, attempted to kill him by inviting him to look out of the train window when I knew another train to be approaching.

This assassination attempt is not something I'm proud of and the only reason I mention it here is because it caused further unpleasantness when, in all the excitement, we missed our station and I had to ask my father to pick us up from Swindon. He missed an episode of *Upstairs, Downstairs* because of us, which featured an actress he found rather alluring. I never heard the end of that and many years later ended up buying him the entire series on DVD just to try and get some closure on the whole incident. When the Bridge called his father to pick him up, he was told that if he was on a walking holiday he should walk home which, to his credit, he did. Had I been old enough to drive, I would have taken him (at least to the bridge).

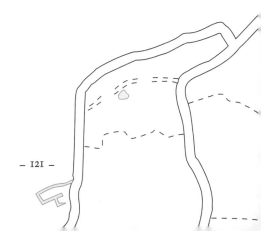

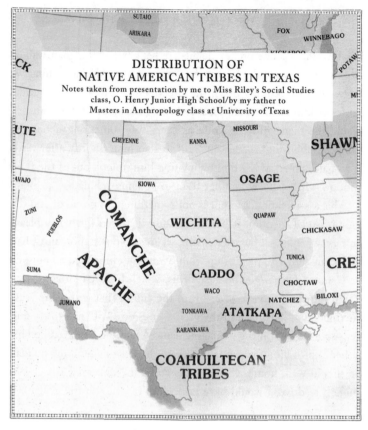

DISTRIBUTION OF
NATIVE AMERICAN TRIBES IN TEXAS

Notes taken from presentation by me to Miss Riley's Social Studies class, O. Henry Junior High School/by my father to Masters in Anthropology class at University of Texas

Atatkapa – named from the Choctaw meaning 'people eater'.
The French colonists eventually put them off the habit.

Tonkawa – great hunters of buffalo. Women vindictive in disposition.
Or dat position. In fact pretty much any position.

Coahuiltecan – prosperous and friendly people. Sophisticated agricultural economy until wiped out by Spanish-borne diseases such as the debilitating *Siesta*.

Karankawa – meaning 'dog-raisers' after their fondness for keeping a coyote-like species. Seasonally migrated between the mainland and the barrier islands of Gulf Coast. Early surfers.

Comanche – first Native Americans to use the horse to hunt buffalo and go on raiding parties. Eschewed side saddle.

Apache – warlike but highly civilised grouping that practised varying degrees of avoidance of the wife's close relatives.

Wichita – They called themselves 'raccoon-eyed people' because of the tattooed marks around their eyes. Also Wichita Linemen because of marks on their bodies.

Jumano – lived in rectangular adobe apartments. Men sported shaved heads with central top knot, style later taken up by Botley punks.

CHAPTER 12

Austin, Texas, USA

I grew up with George W. Bush. We weren't exactly friends but we both had our formative years in Texas in roughly the same place at roughly the same time. If there had been a specially designed programme to help young Englishmen grow up like George W. Bush, then me and the Fatted Calf attended that programme in the years 1976 and 1977. The Sainted One attended a parallel programme designed to create young Laura Bush type women who show immense dignity in the face of acts of unspeakable stupidity executed by close male members of the family.

Texas is very interesting geographically. For a start it is very, very big. It's the largest state in the United States if you don't count Alaska. I didn't like to point it out to any Texans when I was there but that seems like a bit of an admission of failure to me. It's like the Scots saying that Scotland is the largest country in the UK apart from England. But, give it credit, Texas is pretty large. You could fit five entire United Kingdoms within its border. We attempted to do exactly that in the late nineteenth century which is why a lot of famous Texans have British sounding names, such as Bush, Austin, Houston, Johnson and other ones I can't think of at present. At the Alamo, the famous mission station in San Antonio, where Texas fought and lost an important battle in its struggle for independence from Mexico, the names and birthplaces of all the men who died there are listed. They are mostly from England and Wales and a few from Tennessee. When I pointed this out to a Texan standing next to the plaque, he remarked that if they'd had more Texans there, they would have won their war of independence a lot quicker.

Texas also benefits from having a unique shape. If you can imagine a cowboy with a big hat on a horse fording a rough stream painted by a cubist painter such as Braque, then you get the rough outline of Texas. When you get given a jigsaw of the United States with each bit representing a state (I did and I can still name and locate all the states with their capitals), Texas is the first easy bit you reach for. The back portion of the horse is the Gulf Coast, the capital of which used to be Galveston until it was removed from the map, New Orleans-style, by a huge hurricane in 1900. Down south is the Rio Grande, which winds its way up from Brownsville in the east to El Paso in the west, forming a long and spectacularly porous border with Mexico. Jutting up to the north is the elegantly named Panhandle where all the big hats and cattle are. This is where Amarillo is, for all of you who have been asking to be shown the way. It sounds like Armadillo but is the Spanish for 'yellow', largely due to the fact that it's mostly desert.

Being so large, you get five different climates in Texas: mountain, coastal, plains, desert and temperate. I tried to explain that this was impressive in its way but in Oxfordshire we often have four seasons in one day. As some parts of Texas barely have one season in one year, this was often beyond their comprehension. The three industries that dominate Texas are oil, aerospace and cattle. It's a pretty macho set up and it's generally true to say that Texas has outsourced its feminine side to California. Texas is also the only state that has a right to withdraw from the Union by virtue of it being an independent nation from 1836 to 1845. Many of the other states have been pressing it to exercise that option and ride off into the sunset but, as Texans have been running the US for the last few years, it's in no hurry to leave town just yet.

My father decided to take a sabbatical from his teaching job at Oxford University to go and teach at the University of Texas. You may have a mental image of the University of Texas having swing doors and a hitching post to secure your horse while you're matriculatin'. This would be a mistake on your part as the University of Texas is a very fine institution, with colossal endowments and a proud history of scholarship. It's also true that my father was nearly put back on the plane after his first week's teaching when he decided to fail the university's star quarterback for being functionally illiterate. Once this little cultural misunderstanding had been put behind him, he rapidly settled into the university, especially as it had a world-class collection of documents relating to Central American peasantry.

At the same time we children enrolled in our own academic institutions. For the Sainted One it was Cassis Elementary School, named after the pleasant blackcurrant liqueur, and for the Fatted Calf and me it was O. Henry Junior High, named after the famous American writer. I didn't know this when I was there but O. Henry was an American journalist and short-story writer who lived in Austin, Texas, at the end of the nineteenth century. He was famous for his short, pithy pieces written every week in newspapers. While struggling to make it as a writer he worked in the survey office of the Texas Governor making maps. Quite honestly, he might just as well have been me in later life but with a big moustache. O. Henry then worked in a bank, was indicted for fraud and had to flee to Central America where he coined the phrase 'banana republic'. His real name was William Sydney Porter but he used the pen name O. Henry from an abbreviation of Ohio Penitentiary, where he ended up and wrote some of his best stuff. And, incredibly, I am writing this in Belmarsh Prison!*

* Not really. Just stretching the similarities between me and O. Henry too far. Apologies.

O. Henry Junior High was like every American high school you've ever seen in any film. Long, immaculately polished corridors were lined with hundreds of lockers. American high school locker culture deserves a whole chapter on its own. The high school locker is generally the first piece of private lockable property that most Americans own. They are a kind of real-world MySpace lovingly decorated with all the things that make students individuals, i.e. a mixture of religious iconography, rock hagiography and graphic pornography, often in the same locker. That's what makes America great.

Each class looked exactly the same with a number of lines of chairs with integral desk attachments. Bells were rung at the beginning and end of lessons and you were allowed precisely five minutes between each lesson to get to the next. At all other times written permission was needed to be in the corridors. The groups among the kids were also familiar, with the jocks and cheerleaders and nerds and braces wearers. There were also other groups I hadn't come across before: black kids, Hispanic kids, Native Americans and, most bizarre and exotic of all, girls.

Many people in Britain got their first impression of Texas from the hit TV series *Dallas*. Having lived there, I can tell you that the TV version was a watered-down, sanitised version of the reality designed to appeal to more moderate European tastes. The redneck white boys I was at school with wore cowboy boots, belt buckles and, when outdoors, Stetsons. As I learned extremely quickly, this wasn't accessorised by any sense of irony (as I was wearing a hand-knitted jumper and shorts I was probably in no position to make cutting fashion comments anyway). When these mini-rednecks invited you round to play, it wasn't to pretend to fight the Germans, it was to unrack some real weaponry and shoot real live (for the moment) vermin in their back yard/ranch.

The black boys didn't really go for the cowboy look even though a large proportion of the real historical cowboys were black. Instead they'd mostly adopted the urban cool of the hit film *Shaft* and sported huge afros with their metal combs often left in, along with pens, locker keys and other essential items. The Hispanic boys greased their hair and swore at everyone in Spanish. To be honest, they might have been saying 'Welcome to Texas, my funny white friend' but my Spanish wasn't up to much at the time (it still isn't – although I can now swear like a *torro pendejo**).

American schools also have an interesting system by which if you fail a year you then have to retake it. This meant that there were some very large but very stupid people in our class who were probably sixteen or seventeen and were already shaving, drinking and fornicatin' but still had trouble adding up and reading 'Billy-Bob and the Wily Raccoon'. Because the rest of us were all still relatively small, being only twelve or so, it all seemed a little bit like Jurassic Park with an occasional mastodon sitting in the little desk in front of you bellowing unintelligibly.

I shouldn't be too rude because I nearly failed my first month at school. American students carry very large files around with them which give them all the appearance of being very studious. I found out why this was exactly four weeks into my first term at O. Henry when we were asked to hand in our files the following morning. I didn't have a file because I threw away all the work we did as soon as it had been marked on the assumption that no one could possibly want to see that rubbish again. What you were supposed to do was keep all your completed work in a nice file. That night I did a month's work in one night at home and handed in my file in the morning. I got 105 per cent for my file as it was complete and also beautifully bound. Everything was marked out of a hundred at school and I

---

* Bull with burnt knob.

think I got 105 per cent in English for the year I was there simply by having a perfect file and a lovely accent.

I loved pretty much everything about my time at O. Henry Junior High. For a start they'd introduced Latin lessons for the very first time. As I'd already done some Latin in school before I arrived this automatically gave me status equivalent to that of the Ambassador from the Vatican. In Social Studies we were asked to give a presentation on some aspect of local agriculture. My father got wind of this and my subsequent presentation on the local Native American tribes and their preferred method of subsistence prior to colonisation (along with explanatory maps) set new standards of scholarship in the school and across the state and formed the basis for my father's lecture notes for the rest of the term. In my presentation I gave a very balanced account of the troubles the Native American tribes had endured at the hands of white colonists, which seemed to come as quite a shock to the sons and daughters of the colonists and probably to the teacher, Mrs Bottanicus, but made me somewhat of a hero among the Native Americans – well, among Wallace Autavich, the only Native American in my class.

In our music lessons we studied 'American Pie', by Don McLean, which had just come out and was a gift to music teachers across America as it gave them a month's worth of pre-packaged lessons. I can't remember exactly what we learned but something touched me deep inside. Thirty years on I can still sing the full eight-minute version word perfectly and with many of the notes in the right places, although I have to lock the bathroom door to complete the whole song without being physically assaulted. In music we were all encouraged to take up an instrument. As the music teacher knew I wasn't going to be with them for very long, she gave me an 'instrument' that couldn't in fact be played. While she concentrated on the bright kids with the violins, I struggled manfully with an armful of assorted plumbing trying to find somewhere to blow, hit, pluck, slide or bang. I couldn't tell you to this day what 'instrument' I was given to hold but I never managed to get any kind of sound out of it except when I dropped it (mind you, with that kind of training, there's probably a bright future for me in contemporary classical music).

PE lessons were my absolute favourite until I accidentally humiliated the school jock, Alfred Rangel. When I first arrived at the school I had just spent the entire summer on holiday completing epic Long Marches organised by my parents and relaxing after these by running around like a blue-arsed fly. I was therefore as fit as a butcher's dog and, when we all lined up to run the three-mile race, I tore round after the big and beautiful Alfred Rangel and, as they say in Texas, 'whupped his ass'. He didn't take this very well and ordered an immediate rerun over the shorter distance of a hundred yards. This, of course, was playing into my hands as at home this is precisely the distance between a thick ear and being out of earshot and one that I'd been in continual training for over the past twelve years. My standing start was also second to none and I could reach terminal velocity before my mother finished the second word in the phrase 'Come here!' The ass of Alfred Rangel was therefore once again severely whupped and that is how I found myself locked in my own locker for the first time. I missed maths before the caretaker let me out, so all in all I did rather well out of the whole experience.

The other institution we joined in Austin, Texas, was our local church. Not being a member of a church in Texas would be like not having a driving licence or not having a semi-automatic weapon in your bedroom; in other words, madness bordering on outright evil. By a nice coincidence, the church we joined was on the very same crossroads where we witnessed our first shoot-out. We had just cleared the junction when the car behind us was mildly cut up by another car. Naturally enough they both got out of their cars, drew their weapons and started blasting away at each other. The Fatted Calf and I had our noses glued to the back window to see who fell over first.

What happened next I remember very clearly and it still frightens me. My mother was driving and had registered what was happening in her rear-view mirror, which was now blocked by the back of two little heads staring out of the window. In a split second she decided that the best way of distracting us from the carnage was to sing. In the time it takes to draw a Colt .45 she was blasting out that well-known nursery rhyme 'Who Killed Cock Robin?' For us boys, with our eyes glued to the rapidly receding action, it seemed like a rather bizarre Tarantinoesque soundtrack.

The church on that corner (where the fast-driving but slow-shooting motorist was buried) was called the Church of the Good Shepherd. It might as well have been called the Church of the Blessed Donut. I mean no disrespect by this as the church was a devout and well-run place of worship, but there was also a youth club attached to the church where free donuts and sodas were on offer for the price of a quick prayer and one-off circumcision (I'm kidding about the circumcision although those chocolate-iced donuts would have been worth it).

The church youth club was like heaven on earth for a young English boy: there were endless free donuts, large rooms to run about and make a noise in and, God be praised, girls. I was rapidly on the high road to moral torpor and obesity. The youth club seemed to be on a mission to spend money. Every week we had an outing that would have counted as an extraordinary, once-in-a-lifetime extravagance in my family. We went bowling, river rafting, hay riding, witch hunting – you name something extravagant and the youth club organised it. By the end of my time in Texas I was converted heart and soul to extravagance and chocolate-iced donuts.

While we were in Texas, the Fatted Calf was managing his first year of being a teenager and was beginning to worry about his appearance. We'd always worried about his appearance but now he shared our concern. He wanted to buy hip new shirts and jeans and other items of clothing that hadn't been hand-knitted by our grandmother. The trouble was that our pocket money continued to be dispensed with the regularity of pay in the Russian coal-mining industry and required an equal amount of heavy labour.

The Fatted Calf must have been praying for this situation to ease in one of our Sunday School meetings when he noticed the crates of empty soda bottles stacked up by the vending machine. Remember that this was a very popular youth club dispensing free soda and you can imagine how many empties there were. The divine revelation then afforded the Fatted Calf was that the supermarket across the road (called H. E. Butt, believe it or not) paid five cents for the return of empty bottles. From that moment on, no one attended Sunday School more religiously than the Fatted Calf. At least, he attended the first five minutes or so to allow him to stack the crates of empties outside. The rest of the time he spent crossing the road with crates of empties and totting up his earnings. He joined us for a free donut at the end of the morning and we all went home praising the Lord.

George W. Bush and I shared this potent injection of weaponry, sugar and religious fundamentalism in our tender years. For one of us it led on to leadership of the free world, an epic struggle against terrorism and a bloody and unpopular war. For the other it didn't, which just shows that nature is a lot more important than nurture.

SIX FLAGS OVER TEXAS AND MY SIX ALL TIME FAVOURITE FLAGS (NOT INCLUDING THE SUPERB UNION FLAG)

SPAIN 1519-1821
First illegal immigrants.
Brought Catholicism, horses and siestas.

FRANCE 1685-1690
Brought vast range of cheeses and novel way of kissing. Locals left cheeses and ate the French.

MEXICO 1821-1836
Took over Texas just as the British started arriving. General Santa Anna declared himself dictator. Not appreciated by local British.

TEXAS REPUBLIC 1836-1845
War, epidemics, financial crisis. Not a brilliant start. Cowboy career of choice for many Texans.

CONFEDERACY 1861-1865
Texas joins losing side but wins last battle of the war on Texan soil. Cattle now outnumber Texans but remain neutral.

USA 1845-1861, 1865-
Texas becomes 28th state of America and largest apart from Alaska which doesn't really count because it's cold.

CANADA
Maple leaf looks like heraldic design of castle towers. Peru is same flag without the leaf. See what a difference it makes.

ARGENTINA
Smiley sunshine in blue skies which was the weather on their independence day. If the Scots do the same, they'll have the first all grey flag.

MACEDONIA
Lovely warm sunshine on a red background. Makes you want to book a holiday there just by looking at it.

LEBANON
Cedar tree. Difficult to dislike a people who have a tree on their flag.

EUROPE
Organised kleptocracy and bare-faced liberty-snatcher but nice flag.

SOUTH AFRICA
Proof that political sensitivity doesn't mean aesthetic bankruptcy.

CHAPTER 13

Six Flags, Dallas-Fort Worth, Texas

Six Flags is a very large theme park near Dallas, Texas. It's called Six Flags after the six different flags that have flown over Texas in its brief history. Local Native Americans didn't use flags so the first flag that flew over Texas was that of Spain. Various Spanish conquistadores such as Cortés had explored northwards from Mexico but tended to move up the Pacific coast to California as they'd heard the surf was good there. The first Spaniard really to explore Texas was a terrific individual called Álvar Núñez Cabeza de Vaca. His noble-sounding surname actually translates as the rather disappointing Head of a Cow. This name was bestowed on one of his ancestors by the Spanish king of the time, not as a cheap jest at the expense of his bovine-shaped head, but in recognition of his helping the Christian armies defeat the Moors by marking out a secret mountain pass with the head of a cow.

Had I been king I would have given Álvar an honorary title such as 'Lion of the Mountains' or 'Torch of the Andulus' rather than 'Cowhead', which doesn't seem much of a reward for ensuring a major strategic victory. Nevertheless, a couple of hundred years later, his descendant Alvar Cowhead was doing similar sterling work for the Spanish king when he was shipwrecked off the coast of Texas. He spent the next eight years exploring the place and getting to know the local tribespeople. Fortunately for the local tribespeople, there was no gold in Texas and the Spanish soon lost interest.

The next flag raised over Texas was the fleur-de-lis, in a feeble attempt by the French to colonise the Gulf Coast in 1685. In general the French have proved to be pitiful empire builders largely because of the difficulty of transporting soft cheeses great distances. This new Texan venture was no different and foundered after five years of famine, disease, hostile locals and vicious internal bickering over pointless philosophical questions, all consistent with the France we know and love today. Spain took over once more until 1821 when the flag of the newly independent Mexican Republic was unfurled.

Exactly fifteen years later the flag was refurled because Texas decided it didn't want to be Mexican (largely because it was filling up with us British). Texas declared its independence from Mexico and, after some nasty battles such as the Alamo (lost by us British), it, too, became a sovereign nation. Showing an early Texan aptitude for foreign policy it joined the losing Confederate side in the American Civil War (flag five) and was absorbed into the victorious United States (flag six) in 1865.

Of the six flags my favourite is the flag of the Texas Republic, which is still the state flag today. It's very close to the flag of Chile but probably has the stylistic edge. The American flag is, of course, a design classic right up there with our own superb Union flag (this will look decidedly dog-eared if the Scots declare their independence, but then both the Scottish and English flags are world-beaters in their own right. The Welsh have also bagged a pretty decent flag as there aren't many with a dragon in the middle apart from Bhutan, another small and mountainous country with tremendous male-voice choirs). Most modern flags look like different varieties of unpleasant tasting ice cream except for the French flag, which is a corker even though it waves above a land conspicuously lacking in naval victories and bikini area grooming. Forgive my rudeness about the French but this whole Norman invasion thing still rankles.

As you may have sensed already, I've always had a deep affection for maps, and my liking of flags is a subset of this. It's always fascinated me how a rectangle of cloth can carry the affections and respect of a nation. On the other hand, I've got jumpers which have much the same effect on me when I'm wearing them, so maybe emotional attachment to fabric isn't all that strange. It's very important not to be rude about other people's flags because they inspire great loyalty and passion all over the world. If they didn't, no one would bother burning them. You'll notice that we British never get terribly bothered when people start burning our flag, partly because a good percentage of us are already wearing it as underpants. In the illustration at the beginning of this chapter you'll see some flags which I really like. This doesn't necessarily mean that I've been to that country and doesn't mean I am endorsing the political, social and environmental policies of that country (however keen they are to secure that endorsement).

For Americans, the Stars and Stripes is a real unifying force and is displayed all over the place. For example, every school classroom has a little flag in it and every Monday morning all the children in O. Henry had to solemnly pledge allegiance to it. This required standing up, facing the flag, putting your hand on your heart and reciting the pledge which, having heard it a thousand times, I still remember perfectly:

> *I pledge allegiance to the Flag*
> *of the United States of America,*
> *and to the Republic for which it stands,*
> *one Nation under God, indivisible,*
> *with liberty and justice for all.*

As you can imagine, this pledge was a serious problem for me, being, as I am, a loyal subject of Her Majesty Queen Elizabeth II. I didn't want to show any disrespect for America or Americans, who are a generous and kindly people, but neither did I want to pledge allegiance to the Republic just in case the Queen got wind of it and I blew my chance of a knighthood.

Having thought about the correct protocol for quite some time, I decided that the best thing to do would be to stand in a dignified and respectful manner and quietly recite the Cub Scout Promise, which I also remember by heart:

> *I promise that I will do my best,*
> *To do my duty to God and to the Queen,*
> *To help other people [especially Americans]*
> *and to keep the Cub Scout Law.*

This promise covered loyalty to the Queen but also incorporated bonds of international brotherhood so that my American hosts wouldn't feel excluded or snubbed in any way. As I had standing next to me the six-foot seventeen-stone Jim Hickman, who had failed every year for the past seven years, pledging allegiance to the flag at a volume that would have kept it happily fluttering, my Cub Scout pledge would have gone completely unnoticed. I would have been happy for my little compromise to stay unnoticed but unfortunately Mrs Bottanicus, the colossal Social Studies teacher, noticed that my hand was not on my heart.

In America they don't mind where you come from as long as you're basically American: Irish American, Polish American, Iraqi American, they're all fine as long as you have the American bit at the end. What they find difficult to comprehend is that you wouldn't want to have the American bit at all and actually prefer to be Polish Polish or English English. For them your first nationality is a temporary holding pattern until you get called up to the big American melting pot. I foolishly, and probably quite rudely, highlighted my own contrary point of view by using one of Uncle Clive's favourite lines when dealing with Americans he met in his home town of Southend, in Essex (mercifully, for them, quite few). I told Mrs Bottanicus and the whole class that we celebrated their Independence Day as our Thanksgiving Day because we were so pleased to get rid of them. I've said many things I regret in life but that was one of the top three.[*]

Anyway, back to Six Flags and the big theme park. I think Six Flags claimed to be bigger and better than any other theme park on the planet. But, then, claiming something is bigger and better than anything else in the world when you're in Texas is the equivalent of saying 'regular' anywhere else. I'm not very into theme parks because I tend to throw up if I turn my head too quickly at a zebra crossing. The idea of turning myself upside down and corkscrewing at the same time would be tantamount to paying to coat myself in my own vomit. I wouldn't even do that for a concessionary fare. For other normal people, Six Flags was an incredibly exciting and popular attraction with amazing rides such as Old Sparky, Lethal Injection and The Hood. Hold on a minute. I might be confusing the rides with methods of applying the death penalty in Texas, which are equally numerous and exciting unless you're the one person selected for the ride.

The star attraction at Six Flags was a roller-coaster called The Big Bend. This is named after a big bend in the Rio Grande River which forms the southern border of Texas and of the Big Bend National Park, a kind of cactus sanctuary the size of Denmark. For all I know The Big Bend might also be another way of reducing the waiting list on Death Row, although I dread to think what it would

---

[*] If you think I'm telling you the top two, you're badly mistaken (as I was when I said them).

involve. Every year our school arranged a trip to Six Flags. There were ten big air-conditioned Greyhound coaches which meant that pretty much all the school went, unless you couldn't afford it, of course. My parents could afford it but thought it was a total and utter waste of money. While there were lizards on the ceiling at home, what need did we children have of roller-coasters?*

The Fatted Calf took a very dim view of this. By this stage he had hit puberty like a Rock Island diesel locomotive hitting a Datsun Cherry on an unlit level crossing. Coachloads of young Texan cuties were going on that trip and he was going with them if it meant strapping himself to the underside of Mrs Bottanicus, the colossal Social Studies teacher. Then fate, as is its wont, intervened. The Eighth Grade had an annual competition where the student who raised the most money for the school band would get a free trip to Six Flags. There was even a fundraising vehicle laid on: the school supplied chocolate bars, which were then sold at a stiff mark-up, allowing a hefty profit to go to the band.

When the fundraising idea was announced I should imagine there were one or two young Texans who thought they would apply their native commercial instinct to the project and the bus pass would be sitting in the back pocket of their boot-cut jeans before you could say Enron. Clearly they had no idea that the gaping maw of the Fatted Calf's bloodlust for profit was opening up behind them. As soon as we heard about the chocolate scheme, we knew the Fatted Calf would be pocketing the free bus pass and going to Six Flags with his coachload of braces-wearing trainee cheerleaders. My father,

* To my mother's consternation there were indeed small, coloured geckos on our ceilings at home. This was too much like a return to El Salvador to be tolerated so she put a stop to it by spraying the entire house with something chemically equivalent to Agent Orange which cleared the house of all living things, including us, for a week.

who had a very good understanding not only of the Fatted Calf but also of macro-economic and social trends, predicted that after my brother had finished his chocolate sales push there would be an obesity problem in the Austin area.

At first the Fatted Calf turned his attention to the low-hanging fruit, as they say in business. By this I mean he attempted to sell about three hundred tons of chocolate to his immediate family. My mother may have bought a bar to give him encouragement (that's like encouraging a Saturn V rocket by giving it a squirt of lighter fuel on blast off). I should imagine my father also took a couple of bars, taking care to dock the cost from my brother's pocket money, which was now about £3000 in the red. Of course, the Fatted Calf tried harder with me because he knew I was a soft touch and because he knew I liked chocolate. After about ten bars I explained to him that there were limits to how much chocolate a single human being could consume. He, in turn, explained how *foie gras* was made and that I could do a lot better if I tried a little harder.

Pretty soon, the Fatted Calf got into sales overdrive and was selling chocolate to just about every person he met during the day. Stock began to be an issue and he started a little commodity trading on the side where he would buy chocolate from other Eighth Graders who had taken the stock but couldn't make the sale. During the weekends he set up a small stall outside the local supermarket with a range of heart-rending advertising boards focusing on the cherished liberties enshrined in the Texas Constitution and how they would be at risk unless all right-minded Texans bought twice their body weight in chocolate. It won't surprise you to know that the Fatted Calf won his free trip to Six Flags, nor will you be surprised to know that, thanks to his sales push, the school band ended up with an average of four instruments per band member and extra staff to turn the music pages during concerts.

What may surprise you, though, was that I also earned a free ticket to Six Flags. No one remembers this in my family because it runs so contrary to the accepted myths that formed the bedrock of the Browning household. I watched my brother selling chocolate with interest and a certain amount of sickness, and when the Junior Classical League of America, O. Henry Chapter, Seventh Grade, mentioned that they, too, needed to raise funds, I quietly suggested that they might also like to get students to sell something in a competition. Mrs Wilson thought this was an excellent idea[*] and special pens were sourced to be sold.

Not long after, I received my allocation of fundraising pens. I don't have the Fatted Calf's extraordinary sales skills so I simply went to the school office and asked them how much they were paying for the pens they sold to the students. This turned out to be more than I was charging so I shifted my entire stock and that of most of the other members of the Junior Classical League of America right there and then and claimed my free trip to Six Flags. I've never had the recognition this amazing business coup warranted so it's nice to see it in print. A sad consequence of this was that Mrs Wilson never asked me to say anything again; I suspect she thought what I'd done was in some way anti-competitive, bordering on the Communist, and not the sort of behaviour to be expected from a young English gentleman.

[*] Mrs Wilson thought everything I said was an excellent idea. To be honest, she thought everything I said was excellent simply because of my accent. Most British people get about a 20 per cent dividend on their personality when they visit America because their accent is seen as cute. In Texas they hadn't seen many Brits since our help for their cotton industry during the American Civil War, so a British accent was a rare treat for them. I remember having to stand up and say 'the enchanted wood' three times in Mrs Wilson's class because she thought it was so, well, enchanting. To be honest, I played on this a bit and ended up leaving Texas sounding more like Bertie Wooster than George Bush.

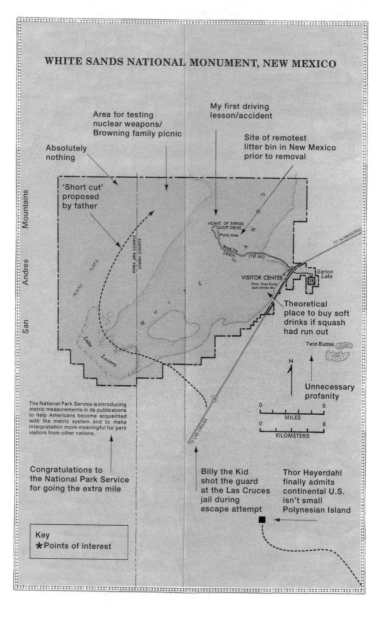

WHITE SANDS NATIONAL MONUMENT, NEW MEXICO

My first driving
lesson/accident

Area for testing
nuclear weapons/
Browning family picnic

Site of remotest
litter bin in New Mexico
prior to removal

Absolutely
nothing

'Short cut'
proposed
by father

San Andres Mountains

HEART OF SANDS
LOOP DRIVE
Picnic Area

Round Trip
26Km   (16 mi)

TO ALAMOGORDO

DOÑA ANA COUNTY
OTERO COUNTY

ALKALI FLATS

VISITOR CENTER
Films Post Cards
Soft Drinks etc.

Garton
Lake

Theoretical
place to buy soft
drinks if squash
had run out

W H I T E

Lake
Lucero

Twin Buttes

N

Unnecessary
profanity

The National Park Service is introducing
metric measurements in its publications
to help Americans become acquainted
with the metric system and to make
interpretation more meaningful for park
visitors from other nations.

70

TO LAS CRUCES

0   5
MILES

0   8
KILOMETERS

Congratulations to
the National Park Service
for going the extra mile

Billy the Kid
shot the guard
at the Las Cruces
jail during
escape attempt

Thor Heyerdahl
finally admits
continental U.S.
isn't small
Polynesian Island

Key
★ Points of interest

CHAPTER 14

New Mexico, USA

Towards the end of our time in Texas it was ordained that we should have a holiday. As most of our time in Texas felt like a holiday, this seemed slightly odd. Nevertheless, it was decided that we should have a holiday while we were there in case we started wanting one when we got back and Kidwelly reared its ugly, soot-covered head again. After some discussion (not including us children, obviously) it was decided that we would drive to New Mexico. This came as quite a relief to us as it was quite within the realms of possibility that the trip would have been classified as an extra Long March.

I've tried to avoid mentioning it so far, but it's now going to be impossible to avoid the subject of our American car. In Kidlington, you may remember, we had two halves of two different cars and we put this purchase down to sheer gratitude on my father's part for his neighbour's help with the burst water pipe. I've thought long and hard but I can think of no plumbing event sufficiently catastrophic that would explain the purchase of our American car.

It's impossible to live in America without a car unless you live in downtown Manhattan. Actually, that's not quite true. A close friend of my father's who lived opposite us in Harwell Hill and who was also an Oxford academic (he was one of the four mega-intellects that needed Langton Machoko's help getting the desk upstairs) visited Austin the year before we did and decided that if a bicycle was fine in Oxford it would be fine in Texas, too. On his first day cycling to work down Interstate 35, an eight-lane motorway, he was knocked off his bike probably by a motorist totally in shock at seeing a mad Oxford professor weaving along the hard shoulder. Fortunately he survived but spent most of his sabbatical in traction. In the light of this, my father decided that a car might be a good idea and made his way to the local car showroom.

American cars in the mid-seventies were absolutely magnificent. They were confidence on four wheels, they were all the size of grit lorries and they all had interiors bigger and more comfortable than my parents' bedroom in Botley. I once had a road trip in a car with the church youth club and several of us young pilgrims were in the back of an Oldsmobile Custom Cruiser, which was the biggest car ever built. The one we were in was the station wagon (estate version), which was even bigger. I remember me and Amy Talbot getting to know each other extremely well in the far end of the car while her parents sat oblivious in the enormous bench seat way up front. Now I think of it, they may also have been getting to know each other extremely well way up front and we wouldn't have known anything about it either. You could do a lot of things in an American car, especially as you don't really have to pay attention to the road because, in Texas at least, they are very big, very straight, and you can veer off them and not hit anything for a hundred miles.

The one thing American cars of that era had in common was that they were vast. The other thing they had in common was that for some unknown reason, possibly to do with the baffling American love affair with the French, they all had Gallic sounding names. We could have had a Chrysler LeBaron, a Cadillac DeVille, a Chevrolet Bel Air, a Buick LeSabre, a Pontiac Grand Ville or a Mercury Marquis with an optional d'Elegance cloth interior. My preference would have been the Ford Country Squire as it was the only Anglophone car on the market.

We could have had any of these cars until my father mentioned his available budget. We were then hustled out of the showroom and taken for a long walk to the far end of a dusty parking lot. On the other side of the concrete fence was a breakers' yard. The salesman, whose charm factor had slipped considerably since mention of the budget, pointed to a car that was sitting forlornly as close as it was possible to get to the breakers' yard without actually being in it. I'll remember that car for as long as I live. It was like the donkey they use in advertisements for donkey sanctuaries. Cleary it had been horribly misused and, to be honest, wasn't very attractive in the first place.

It was called a Ford Maverick. They'd obviously had the British Leyland marketing team over for that one. They must have looked at it and said it's a bit of an ugly duckling. This would have eventually been converted through marketing alchemy into 'Rugged Loner' and then on to 'Maverick' (this was long before Tom Cruise sanctified the name 'Maverick' in *Top Gun*). Whatever it was called, it was plug ugly and deserved to be put down humanely. They were probably waiting to heave it over the fence and do exactly that when our little shiny-faced family with the micro-budget stepped into the showroom. Our particular Ford Maverick had cellulite. Down the rear flanks, the metal was dimpled and rough, as if it had been crudely hammered back into shape after several nasty incidents, which it had. Cleverly, the car had been repainted an attractive rust colour. As least I'm assuming that was paint.

My mother, who tends to be more practical in these circumstances, wisely asked the salesman if the Ford Maverick had air conditioning. The salesman shot a wad of tobacco across the lot, smiled like a rip in a wet paper bag and assured us that the Maverick was blessed with the latest 4/70 air conditioning. Money exchanged hands and we drove out into the wide Austin highways proud and happy until five hundred yards down the road where an oncoming truck reminded my father in no uncertain fashion about the quaint American tradition of driving on the wrong side of the road. There's a cowboy song called 'Hell in Texas' which has a line 'The heat in the summer's a hundred and ten, too hot for the devil and too hot for men'. We bought the Maverick in mid-August and by the time we got it home we realised that there was no air conditioning whatsoever in the car. My father phoned the car showroom and asked about the 4/70 system. They told him that it was four windows open at 70 mph. Shortly after this exchange my father's university teaching took on a marked anti-capitalist slant.

This was the car in which we were to drive to New Mexico on holiday. Texas has a moderate climate compared to New Mexico. It should also be remembered that we three children were still mostly wearing shorts so our little bare legs were in direct contact with seat plastic that reached temperatures that could flash-fry beef. Fortunately, having worn shorts all our lives in all temperatures we no longer had any nerve endings in the skin on our legs, and to that fact alone I attribute our survival on the long drive to New Mexico.

Mind you, it nearly killed us in other ways. Somewhere along a remote back road across the New Mexican desert – and remember the remote back road has always been my father's road of choice – we got a puncture. I'm amazed we got that far without having a puncture because, when the car was purchased, the tyres had a rubber coating no thicker or more securely applied than the sticky surface of a toffee apple. Once we'd stopped, my mother ordered us to stay in the car to avoid being run over by the speeding traffic that might suddenly appear between us and the horizon. So we jumped around on the back seat and wondered why our father was getting increasingly grumpy as he tried to jack the car up. About an hour later the tyre still hadn't been fixed, possibly because my father was absolutely convinced that his new way of removing the tyre would be more efficient and painless than that recommended in the car handbook my mother was simultaneously quoting from at great length and to absolutely zero effect.

As my father struggled with the tyre and I watched small grains of sand blow across the road, I recalled a story about the New Mexican desert I'd read as part of my preparatory homework for the trip.[*] The early European settlers had crossed the desert in their wagon trains looking for an easy way to reach California. Sadly they had picked just about the most difficult way and a fair few of them perished in the middle of the endless desert sands. The desert did what deserts do, and wind-blown sand eventually covered the remains of these wagon trains. A hundred years later, when the sands eventually moved on, the perfectly preserved remains of the early settlers, still huddled in their wagons, were revealed. Sitting by the road, I imagined a hundred years on when the remains of the Browning family would emerge from the sands, my mother still with the handbook in her hand and my father desperately clutching a tyre lever in a totally ineffective manner. The guide would hazard a guess that the family perished because of total inability to change the tyre and would marvel at the shorts worn by the poor dead children.

Once I had completed this day-dreaming I decided it was time to water the desert and I wandered out into the scrub by the side of the road. I had gone no more than twenty yards or so when a passing pick-up truck braked sharply and parked right behind our car. We assumed he'd come to laugh at our Ford Maverick but the driver was actually more concerned that we were about to become rattlesnake lunch. They wear cowboy boots in that part of New

---

[*] There was always preparatory reading to be done before our holidays, otherwise we wouldn't know anything about where we were going and might miss some point of historical, geographical or geological interest.

Mexico to stop themselves being bitten by rattlesnakes and our little bare legs must have looked like a starter of spring rolls to the hundred thousand rattlesnakes that were sunning themselves in the immediate vicinity. Having shepherded us back to the road, the Good Samaritan drove off with a cheery wave, but not before he bent down and quickly changed the tyre using a little-known method very similar to the one in the handbook.

Four hundred miles to the west was our holiday destination, Cloudcroft, a little village in the mountains overlooking the White Sands Desert in the southern half of New Mexico. Cloudcroft sounds like a lovely place to stay. That's not altogether surprising because it was given that name by the early settlers who wanted to promote tourism to the area. If you had the choice, what would you call Blackpool if you wanted to attract visitors to the area? Or Cleethorpes, or, for goodness' sake, Skegness? I think Cancun and Phuket need some attention as well. On the other hand, I don't need to be told twice to go to Copacabana, Barbados or the Seychelles because they sound beautiful long before you get there. Cloudcroft was a lovely place, eight thousand feet up the mountainside overlooking a glaring white sand sea. Unfortunately my father had repeated an elementary mistake and borrowed a cottage from an academic colleague. I don't recall many details about our accommodation in Cloudcroft principally because something that hasn't been fully constructed doesn't have much detailing.

On our way to Cloudcroft my father had selected for our delight a site of special geological interest, the White Sands National Monument. Highway 70, which took us to the White Sands National Park, also ran through the White Sands Missile Range. There were notices every so often warning us that the highway would be closed for test firings of missiles. We were also near the Trinity site, where the first atomic bomb was tested. There was palpable tension in the car as my father desperately wanted to veer off the highway and go and see for himself what test firings of missiles entailed and whether they did indeed constitute a danger to passing pedestrians. I'm only glad we weren't there thirty-five years earlier when they were testing the atomic bomb as I'm sure my father would have found a way of breaching security and getting us all a ringside seat.

The White Sands National Monument is approximately four hundred square miles of entirely featureless, flat sands totally uninhabited and largely unloved. This is where my father decided to let me have my first go behind the wheel of a real, full-size car. When I said entirely featureless I forgot to mention the one litter bin in the desert carefully put there to keep those four hundred square miles clean. In my short time behind the wheel I managed to make the desert even tidier by removing the litter bin. It wasn't entirely my fault as I couldn't really see out of the front windscreen apart from the line of hazy blue mountains some two hundred miles to the north. Also, it's tricky to pick out small features like litter bins when you're travelling close to 100 mph with your mother screaming at the top of her lungs right behind you. There may also have been damage to the car but, given the state the car was in when we bought it, we couldn't be sure.

We had a lovely holiday in Cloudcroft. At least, I have no memories of anything traumatic happening. On the long drive back, I renewed my close relationship with verges when we stopped somewhere remote to do something we couldn't practically accomplish in the car. There was featureless scrub as far as the eye could see in each direction and yet I found myself looking closely at a weathered old wooden fencepost and the rusty remains of some barbed wire. It struck me then how little impression on the landscape humans had made in that part of the world. In Botley and most of the rest of England the ghosts of previous generations lie multilayered across the landscape like a Walls Viennetta of history. Old Hengist of North Hinksey was a thousand years old and Wayland's Smithy on the Ridgeway south of Oxford was at least a thousand years older than that.

That old fencepost in the middle of nowhere was the kind of spot you never thought you'd find yourself in and to which you certainly never thought you'd return. Strangely, though, the Sainted One did return. At roughly the same time as my future girlfriend (see Chapter 15) was chaining herself to the railings of Greenham Common airbase and bringing the Cold War to an abrupt conclusion, my future brother-in-law was landing American bombers on the tarmac within. The Sainted One, when she was still a teenager, met this dashing American pilot and love hit them like laser-guided ordnance. They were swiftly married and eventually the pilot received notice of his next posting. His particular plane was stationed in three places in the American empire: Oxfordshire, California or New Mexico. Once fate takes you to a place it often takes you back, maybe through sheer laziness on the part of fate or a shortage of capacity in the destiny travel department. Whatever the reason, the Sainted One found herself a few miles down the road from that fencepost scratching a living from an airforce base the size, complexity and general attractiveness of Swindon.

That trip to New Mexico was notable for the very few things that were notable, perhaps not surprising in a desert. We had to make the most of small moments of excitement and one of these came in the form of a spontaneous witticism from my father that occurred in a very remote convenience store in a tiny, one-horse town in West Texas called Muleshoe. It has been preserved for posterity not only by my father's repetition of it, but by my Uncle Clive's adoption of it as part of his one man Offend-America programme.

FATHER: 'I'd like some bread and orange juice.'
CASHIER: 'You're not from round here are you?'
FATHER: 'No, I'm from England.'
CASHIER: 'Wow! Why have you come all this way?'
FATHER: 'To get some bread and orange juice.'

We all derived a great deal of pleasure from this exchange especially as it meant a reprieve from the normal repertoire of 'college' jokes. College jokes arise in the staffroom of any academic institution to fill the gaping void between academics when they attempt two-way communication as opposed to one-way lecturing. Every academic is such an expert in their field that it would take at least ten years' preparatory study to exchange even the smallest pleasantries in their particular area. Instead, old, carefully crafted and totally innocuous jokes are passed round the table like the port; savoured and appreciated and leaving everyone feeling warm inside. These jokes don't survive contact with the outside world, but mercifully academics don't have to go there. Some college jokes are so old that they are referred to simply by their punchline. For example, we grew up knowing that 'but Lord he had a hat' was the punchline of a revered and ancient college joke without ever knowing what the preceding story was, if indeed there was one.

The formal introduction/warning of an impending college joke was generally the words 'Which reminds me of the story . . .' The joke is then told while those listening assume the pre-punchline brace position of amused and excited attention, ready to break into laughter at the payoff. This position has to be assumed immediately as there is no indication when the punchline has arrived except for the fact that the next line is always 'which I thought was incredibly funny/rather amusing/exceptionally entertaining'. Actual laughter isn't required at this point, rather a communal nodding of heads to acknowledge that a joke has passed among those assembled. Verdicts along the lines of 'that really is very, very funny' are then delivered as if sentencing a minor offender to three months' community service. Occasionally someone will then make the mistake of saying, 'that reminds me of the story' and everyone will have to assume the brace position again. It's nice to know that there are probably academics at Oxford to this day who are saying, 'bread and orange juice' to each other and laughing politely without a clue what it means or where it came from.

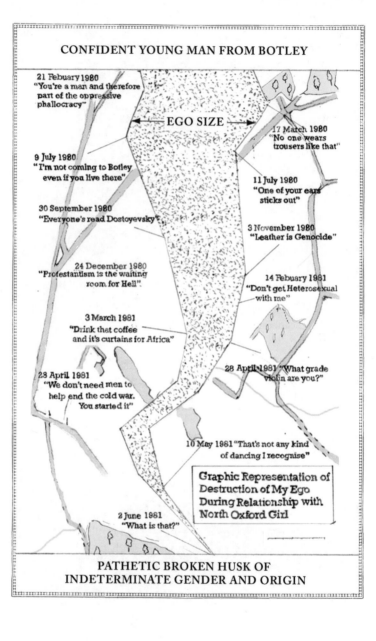

CONFIDENT YOUNG MAN FROM BOTLEY

21 Febuary 1980
"You're a man and therefore part of the oppressive phallocracy"

←— **EGO SIZE** —→

17 March 1980
"No one wears trousers like that"

9 July 1980
"I'm not coming to Botley even if you live there"

11 July 1980
"One of your ears sticks out"

30 September 1980
"Everyone's read Dostoyevsky"

3 November 1980
"Leather is Genocide"

24 December 1980
"Protestantism is the waiting room for Hell".

14 Febuary 1981
"Don't get Heterosexual with me"

3 March 1981
"Drink that coffee and it's curtains for Africa"

28 April 1981 "What grade violin are you?"

23 April 1981
"We don't need men to help end the cold war. You started it"

10 May 1981 "That's not any kind of dancing I recognise"

Graphic Representation of Destruction of My Ego During Relationship with North Oxford Girl

2 June 1981
"What is that?"

PATHETIC BROKEN HUSK OF INDETERMINATE GENDER AND ORIGIN

CHAPTER 15

North Oxford, Oxford, England

Most cities have a particular part of them with a distinct, slightly stronger flavour, like a rogue caper on a Pizza Margherita. In Oxford this is an area called North Oxford in, believe it or not, the north of the city. If you want to find out what big idea will be doing untold damage to the fabric of British society thirty years from now, simply tune in to the deeply held beliefs of the current North Oxford set. North Oxford is where the university's academics live and the definition of an academic is someone who knows more and more about less and less. At work the academics are safely confined within their laboratories or libraries, but after work they can let their intellects loose on the wider problems of society about which they know a lot less than people half as clever as they are. It was through this very special set of circumstance – i.e. people of great authority talking with great authority about things they knew precisely nothing about – that Oxford became known as the Home of Lost Causes.

The girls who grow up in North Oxford are a breed apart. If you aren't trilingual by four, with grade seventeen on the violin, you are regarded as having special needs. Similarly, if you have a first name that is shared by more than two other girls on the planet you are almost invisible. The paint shop for these glistening intellectual machines is the Oxford Girls School. This fine school doesn't believe in female equality, it believes in the total supremacy of women in all departments and views men as little more than moist wipes. As you can imagine, going out with a girl from this school is a bit of a challenge, to say the least. I wouldn't have dared suggest it but I had one literally thrust upon me. I was involved in a play at my boys' school and for this breakthrough in dramatic theatre we needed real, live girls. Three girls volunteered to come across from the Oxford Girls School. If OGS girls were scary in the first place, I should have guessed that the ones who volunteered to come into the heart of male darkness would be their elite storm troopers, and so it proved.

One of the three (all of whom wore nothing but black – you can imagine the overall effect) decided she wanted to have me as her boyfriend. I can't remember any good reason for this but she must have had some hidden charitable/research purpose. I do remember our first date, though, which was a walk in University Parks. It took us about three hours to cover five hundred yards, which I found extremely difficult as I would normally have been expected to cover seventeen miles in that time on a family outing. In those three long hours I was introduced to Feminism, Marxism, Pacifism, Existentialism, Socialism, Literary Criticism and Roman Catholicism. As her mouth was working at the pace I normally walk, it was hellishly tricky to get a kiss started. But then, as I soon learned to my cost, a kiss which in Botley is viewed as a pleasurable activity done with the mouth was viewed by my North Oxford 'girlfriend' as a political act bordering on a declaration of war.

As homework before my first date I was given a copy of *The Female Eunuch* by Germaine Greer.[*] For those of you who haven't read it or seen the excellent DVD version with Reese Witherspoon, it basically catalogues the thousands of years of male oppression of

---

[*] Women seem to make a habit of giving me a reading list before I go out with them. I once went out with a woman in medical publishing. On our second date she brought along one of her publications on erectile dysfunction. From then on I didn't know whether I was a boyfriend or a research project.

the female. The obvious if unspoken subtext was that our date was going to be an attempt to write an additional epilogue to this work where men (represented by me) finally get a thousand years' worth of comeuppance in one go. Which wasn't quite what I'd been hoping for when I agreed to have the date in the first place.

We went out for nearly two years. The relationship deepened and strengthened. To be accurate, I deepened and she strengthened: I started to read books that weren't about the Second World War (there are a few here and there) and she was able to walk up to three miles at a reasonable pace. This strengthening on her part sadly led to one of the most traumatic elements of our time together: traumatic for me, that is, as her own resting state was profound emotional, political and aesthetic trauma.

One lovely evening we went for a long (in her terms) walk along the banks of the Thames as it winds through Port Meadow, north of Oxford, between the Perch Inn and the Trout Inn.* When we arrived at the Trout she realised she'd lost an earring and was quite upset about it. I was quietly relieved as it meant a break from her being upset at me for some heinous ideological transgression. We retraced our steps but failed to find the fiddly little thing amidst the long grass and cowpats. Eventually she went home still upset but I went home with a plan.

I loved that girl in my sexist, patronising, oppressive, smug, middle-class, unfeeling, bourgeois, heathen, Anglo-Saxon, oafish way, and I decided I would find her missing earring. I explained this idea to my mother and she, to her eternal credit, immediately volunteered to search with me the following morning, explaining that needles in haystacks, their location and extraction thereof, were a speciality of hers. For his part my father immediately volunteered to go to work

---

* There should be a pub called the Happy Angler in between but, sadly, there isn't.

as usual as he had never heard such a foolish idea in all his life (I'm
sure he had – our family is famous for them). At dawn the following
morning, my mother and I retraced once more the steps of the night
before. We retraced them and then we retraced our retracing. Finally,
as the hours slipped away, we retraced them one final time and there,
at roughly the spot I'd announced a sneaking regard for Mrs Thatcher,
prompting some kind of spontaneous war dance by my 'girlfriend',
we found her earring.

That night I presented it to her and told her that I'd retraced her
steps a thousand times before I found it. I didn't say my mother had
been involved in the search, as somehow that didn't really add to the
romance of the gesture. What happened next was the traumatic bit.
She didn't believe me. She thought I was making the whole thing
up. She thought I'd just pocketed the earring after it fell off and had
been teasing her ever since. No amount of persuasion could shift her
in this belief. I even felt like calling my mother in as a witness but,
like I say, if that had persuaded her it would at the same time have
removed much of the romantic gloss. It occurred to me afterwards
that my 'girlfriend' must think I was capable of this kind of under-
hand behaviour and then, shortly afterwards, it occurred to me that
I probably was.

A few nights previously she'd been on the phone talking to me
at great length about an obscure poet I'd never heard of – Wordsworth,
I believe – and berating me for being vulgar and poorly read. While
she was banging on this particular drum, I put the phone down
gently on the table, crossed the room and selected *The Oxford Book
of English Poetry*. Fortunately, someone had taken it out of its wrapper
and I soon found the poem in question. I picked up the phone which
was still vibrating with indignation and then proceeded to quote big
chunks of the poem back to her 'from memory', taking care to get
a few bits wrong to sound extra authentic. I told her the poem had

long been one of my favourites. For a few seconds my star burnt brightly in her firmament.

Going out with the OGS girl was a hugely educational affair and I wouldn't have swapped it for anything, apart from possibly a nice girl who could kiss well. That said, once you've been out with a North Oxford girl, for the rest of your life you never fear any other woman, man or, for that matter, wild beast. If she was an education, having tea with her family was a doctorate. Her father was one of the cleverest men in the world. Most academics think they are the cleverest people alive, but he really was. He was another Oxford don (you couldn't throw a brick without hitting one in North Oxford). His speciality was legal ethics which may seem like a contradiction in terms to most people but certainly wasn't to him. He was a deeply, deeply moral man and a devout Catholic. I know that Catholics only come in two varieties, devout and lapsed, but he was truly devout and was partially responsible for defining the nature of Catholic devotion as, according to his daughter, he helped the Vatican with tricky points of theology. Mind you, according to me, my mother is descended from Anne of Cleves, so it may not be gospel.

The father of my girlfriend was a very quiet, unassuming man and he used to sit very still with his brain humming like a nuclear power plant, making very little noise but generating great power. Humour was not really his thing, and as sarcasm and foolishness were also out, it meant that we had precious little means of communication between us. Nevertheless, he was an attentive listener and gave great weight to my most lightweight comments, which were the vast preponderance of all the things I said. Normally I would make a little comment and there would be a long pause as my minor witticism passed down the long marbled corridors of his intellect. Very occasionally there would be a puff of white smoke from his Baroque cerebellum in the form of the thinnest of smiles. But mostly everything I said was greeted with the knowing smile of a father confessor hearing once again an act of idiotic banality.

His wife was equally interesting. She struck me as actually being the more lively of the two but she had a permanent air of having her train of thought interrupted. This was principally because she was always having children, the result of new and advanced forms of birth control they were crash-testing for the Vatican. She tended to encourage me in my attempts at humour as she saw this as a way of mildly torturing her husband who was too polite simply to excommunicate me and hand me over to the Spanish Inquisition.

Educational though my relationship with the North Oxford girl and her family certainly was, I decided it would have to end when she tried to kill me during the storm scene of *King Lear*. Our breaking up had been coming for some time. She had tried a number of little playful attempts to get rid of me but then aborted when I didn't seem to mind enough. Now that our relationship had settled down once more, the stage was set for a really major bust-up. As part of her relentless tailor-made programme of intellectual and cultural uplift for me, she had arranged for me to come with her family to see *King Lear* at Stratford-upon-Avon. She'd made it clear that *King Lear* would be particularly instructive for me because it demonstrated how foolish men are, especially members of the oppressive ruling classes, and how dreadfully they treat women, especially their own family members. Everything passed off peacefully until the interval. We went outside and had some trivial argument, possibly about ice cream. Taking a leaf from *King Lear*, I decided to punish her for her disobedience and ignored her.

My girlfriend liked arguments and was never happier than when sharpening her intellect on the whetstone of political or philosophical debate. What she couldn't stand was being ignored so quite rightly she decided to kill me. This decision was slightly mistimed as the interval was over and we were now going back into the theatre. I sat down between her father and her, the lights went down and King Lear came back looking nearly as angry and full of hatred as my girlfriend, who was now bubbling up into some kind of thermonuclear apocalypse beside me.

I was very conscious of sitting next to her father (who I quite liked) and also that we were in a vital part of the play, so I tried not to respond when she grabbed the hair on one side of my head and started trying to pull it out. It wasn't long before the famous storm scene on the blasted heath was in full spate, with poor old King Lear having his wits lashed by the wind and rain. I, too, had waves of pain passing through my head, as my soon to be ex-girlfriend tried to scalp one half of it. Goodness only knows what the people behind us must have thought: they must have been getting double the entertainment. At the height of the storm she gave my hair one last, colossal yank, which I obviously ignored, and then she got up and stormed out of the theatre taking, I have to say, a large portion of my hair with her and some of the limelight from the poor old king on stage who was still gamely trying to muster a bit of drama at his end.

It was quiet in the car on the way home. We sat in the back with an odd sense of friendliness akin to those strange relationships struck up between people on Death Row and the oddballs who choose to write to them. I was particularly pleasant to her parents in the car because I wanted them to remember me as a nice boy. When I finally got back to my own home in beautiful Botley I sat down and began to write my girlfriend a letter that would in a literary sense kick to kill. I have to admit that I said some rather unpleasant things in that letter but I'm even prouder to admit that I threw it away without sending it. I then wrote her a disgusting sickly-sweet Battenberg cake of a letter where I talked about our different trajectories of growth and how I was blocking her sunshine and how I'd learned so much from her and other such execrable duck-billed platitudes. She wrote back saying she respected me more than she could possibly say and that there would never be another like me. There wasn't. The day after I received the letter she started going out with a Jewish lawyer, possibly because he was an even greater conversion challenge than me.

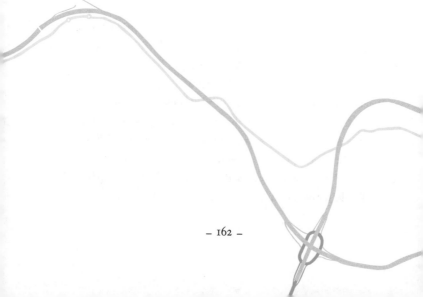

ANDALO, ITALIAN DOLOMITES
WHERE I LEARNT EXTREME SKIING

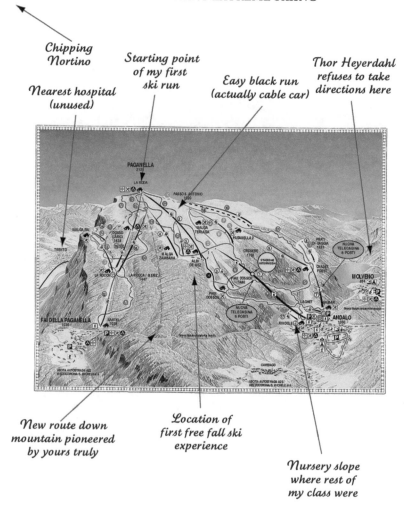

Chipping Nortino

Nearest hospital (unused)

Starting point of my first ski run

Easy black run (actually cable car)

Thor Heyerdahl refuses to take directions here

New route down mountain pioneered by yours truly

Location of first free fall ski experience

Nursery slope where rest of my class were

CHAPTER 16

Andalo, Italy

By the time I was fourteen I knew beyond a shadow of a doubt that I was not marked out for sporting glory or even basic competence. Anything that involved a ball seemed totally to escape me, as did the ball. It was no lack of commitment, willingness or physical fitness on my part; it was just me being a total klutz. If there is a sporting equivalent of dyslexia, I have it. My biggest fear in life is walking through a park, a football accidentally escaping a game and me being required to kick it manfully back. And that, I suppose, is what first attracted me to skiing. Having investigated the sport closely I determined that a ball wasn't involved in any way. All that seemed to be needed was the ability to skid, and any fool could do that. When the opportunity arose to go on a school skiing trip I thought this might be just the sport for me. But first there was the small matter of assembling the right equipment.

When my mother and grandmother are faced with cold they instinctively reach for their knitting needles. Neither really grasped the concept of skiing; to them it seemed like nothing more than an extended opportunity to catch one's death of cold and, what's more, to die in a country devoid of the most basic health and sanitary facilities (Italy). Three weeks after announcing my intention, nonetheless, to go to Italy and ski, the first item came off the knitting production line in the form of a Fair Isle jumper. I have to say that this was a very nice jumper, and beautifully knitted, but my affection for it faded as it went past the belt. This was partly because the jumper went a lot further than the belt, roared past the thighs and came to rest millimetres above the knees. My grandmother knew, perhaps from bitter experience, that getting your kidneys cold was the high road to hypothermia and that any hint of untucking or uncovering around the waist and you might as well throw your kidneys in the deep freeze. With this jumper that eventuality was physically impossible, as was spending a quick penny or, indeed, bending at the waist.

That said, the jumper was a thing of simplicity and elegance compared to my mother's offering in the war against cold. My mother's own particular concern when it came to cold and catching one's death thereof was the knowledge that at least 400 per cent of body warmth escapes through the head. Her vigorous and practical response to this threat was to knit me a balaclava. This incredible piece of knitting had some amazing design features which I soon encountered on trying it on for the first time. The first was that there was nowhere to look out from or, indeed, breathe in through. I got very close to panicking before I found a razor-thin slit for breathing and two tiny bullet holes to see through. Simultaneously I found that I now had trouble moving my arms as my mother's design had ensured that the neck, shoulders and upper torso were also amply protected from the cold. The final unique design feature was that the wool she'd chosen seemed to be a wire wool and sandpaper mix that didn't so much irritate the skin as completely remove it.

There were two final touches to my skiing outfit. I had a superb pair of leather gauntlet mittens with rabbit-fur lining, which could have been made by my grandmother after a standard Sunday lunch but were bought by my father as a souvenir from Seville (don't ask). These gloves were huge, each the size of an adult salmon, and were the kind Captain Scott had secured round his neck by a canvas harness in case he dropped one while having a stoic leak. Given that the hand couldn't bend within them, they were not ideal for holding ski poles but, as they were lovingly packed in my suitcase, none of us was to know that. The final item I assembled for my skiing trip was climbing boots. If my parents knew one thing, they knew that good boots were essential to a holiday, especially if you were covering up to a thousand miles a week on foot. I had an exceptionally good pair and I think I travelled in them, pretending they were a kind of ruggedised après-ski moon boot.

It took most of the first day in the resort for the laughter of my friends to die down when I first tried on my total Antarctic package. I had none of the ski jackets and salopettes that the others had but I could not, after this reception, wear my hand-knitted polar explorers' gear on the slopes. In the end I had to go on the slopes in my casual trousers, light sweater and coat. The effect was to make me look as though I was a very relaxed, very experienced but slightly eccentric skier who knew he could get down and off any mountain long before the weather got nasty. This may have led to some overconfidence on my part. Certainly the time it took me to wriggle out of my bala-clava and knitted body stocking made me miss the meeting for the beginner skiers at the bottom of the nursery slope.

By the time I had collected my skis, strapped them on and trav-elled to the nursery slope by 'herringboning' the equivalent of a good catch by a medium-sized Fraserburgh trawler, the beginners' class had gone and there was no one anywhere I recognised. I then began to make a series of mistakes that, cumulatively, were to put me off skiing for life. My first mistake was consulting a map. Now I have to say that consulting a map is very, very rarely a mistake in life. Maps have saved me from fates worse than death, and death itself, on many occasions. I have also spent nights in with just me, a couple of beers and a box of Ordnance Survey maps for the sheer pleasure of it. But on this one particular occasion consulting a map was a mistake even though it was a particularly nice one (as you can see at the beginning of this chapter).

I looked carefully at all the routes down the mountain. There were orange routes that twisted and jinked all over the mountain, but there were also reassuring black routes that took a sensible straight course directly down the mountain. I thought, why make life difficult for yourself? You haven't learned how to turn yet: take the easy route, take the black route. I then checked out the ski lifts. There was some kind of chair lift nearby but it looked rather tricky to get on to. Instead, I opted for air-conditioned luxury aboard a cable car, wisely saving myself a few million lire by getting the off-peak ticket.

I queued up for the cable car and would probably have got in a lot faster if someone had told me you have to take your skis off beforehand. I tried to wedge myself through this ludicrously small door thinking its designer must have assumed that skiers were either completely double-jointed or had a degree in geometry. Eventually someone got tired of this, stuffed me bodily through the door and threw my skis in after me. The doors closed behind me and up we went. I looked round and there was just the three of us in the car, me and two really professional looking skiers in fancy ski leotards with white lipstick, mirror lenses and skin tanned to a rich-brown parchment. I could see my reflection in their sunglasses, a credit to Britain in my grey V-neck, brown corduroys and duffle coat.

As we began to grind upwards I saw the rest of my classmates lined up at the head of the nursery slope below, clearly having a whale of a time carving this way and that with their fully extended snowploughs. The cable car continued smoothly upwards. Past the first line of trees I could see skiers descending confidently in smooth, curving arcs. It all looked like a lot of fun. We continued our relentless ascent up through the clouds, leaving the valley far behind and finally arriving right at the top of the mountain.

When we finally got out it was bitterly cold, no trees or vegetation, just ice and little stone crosses marking where long-forgotten teams of climbers had perished. The view was magnificent. You could see the Dolomites stretching away on both sides and I fancied that, somewhere in the far-distant valley, I could even see our village. Off the two professional looking skiers went. They were good, sliding expertly away and dropping over what seemed to be a sheer cliff, instantly disappearing from view.

After a good half an hour spent strapping my skis on and losing half my core body temperature, I looked round for the start of the sensible black route and realised that where the two experts had disappeared was the sensible black route. I moved cautiously over to the edge of the slope with my skis in a defensive snowplough braking position. In the glaring sunshine it was difficult to tell where the actual black slope began. Before I knew it I was there, my skis tipped over the edge and I started to experience the thrill of skiing. It was the first time I had ever moved at speed on skis and I immediately assumed the Browning racing position: skis stuck out in front, me sitting down on the backs of them, arms flailing uselessly on each side, eyes wide with terror and gaping mouth filled with a frozen scream.

Within seconds I was approaching terminal velocity. As I rattled down the incredibly steep slope bouncing along on my arse, I had time to reflect that I was about to hurt myself extremely badly. The only consolation was that as long as a tree didn't suddenly throw itself between my skis I'd probably just hit soft snow and be relatively lightly injured. Then, would you believe it, there was a turn in the slope. It wasn't on the map, and it certainly hadn't been advertised or signposted. There were no friendly skiing wardens signalling this way, please. I had an uneasy feeling that I was going to go straight over the edge, drop two thousand feet, impale myself on the

unnecessary wooden spike they put on the gable end of the picturesque alpine chalets in the region, and finally be carried off to the
great après-ski in the sky.

I looked down at my skis to work out how to turn, but they were
absolutely straight. No wheel or hinge or anything. Turning was a
physical impossibility. With a quick prayer to St Melissa, the patron
saint of free fall, I missed the turn and shot over the edge. In strict
trigonometrical terms it was not a vertical drop, but as no strict
trigonometrists happened to be skiing in the vicinity I was more
than happy to call it a vertical drop. By the time I had finished this
little semantic discussion I had dropped a thousand feet. My skis
touched the snow just long enough for me to rapidly reassume the
Browning racing position before again hurtling unimpeded down
through the fresh alpine air.

By now you're probably visualising me head to foot in plaster and
in multiple traction at the local hospital, but it didn't happen that
way, at least not for me. What happened was that I shot straight
back on to the main piste, still perfectly out of control sitting quietly
on the back of my skis. This continued right to the top of the nursery
slope where I finally managed to kill my speed by ploughing straight
through a group of little schoolchildren having one of those unnecessary lessons. I finally came to a stop right at the bottom of the
nursery slope by hitting a large woman carrying a tray of mulled
wine. When you've been thrown from your horse, you've got to get
straight back on otherwise the fear will overtake you, and that's why,
right then and there, I decided I would never ski again for as long
as I lived.

For the rest of the week, I did attend the nursery classes (they'd
cleared the remains of the schoolchildren off the slope by then) but
I never varied from a snowplough position that completely prevented
all forward movement whatsoever. However, I didn't feel good about

being a total loser at skiing and I didn't want to leave Italy with the mountain thinking that it had somehow got the better of me. Therefore, on the last day, when my friends were happily skiing down the mountain, I decided to climb it. This, of course, gave me the chance to wear my full polar exploration kit. I tramped up the hill and, after about fifty yards in my Fair Isle body stocking and sandpaper balaclava, I nearly died of heatstroke.

I climbed all day roughly following the line of the cable car, but not close enough so that anyone could see me too clearly from it. At about three o'clock in the afternoon I realised that I wasn't going to get to the top before the cable car closed. I was also getting very cold. I then made another of those decisions that has probably been the hallmark of great explorers down the centuries: I decided to take a short cut back. Away to my right I had occasionally glimpsed, far beneath me, a road that snaked its way through the valley back to the village. My bold plan was simply to climb down the slightly steeper side of the mountain to the road below and then follow it home.

My earlier descent of the mountain on skis had been frightening but at least it was done at high speed and in relative comfort. Climbing down this side of the mountain was possibly one of the most foolish things I've ever done in my life. The slope was tree-covered and I naïvely assumed that, if a tree could grow there, it wouldn't be that steep otherwise the tree would fall over. I learned over the next half a mile of bottomless drop that the only slope trees don't grow on are overhangs. I didn't climb down that slope; I lowered myself, I slithered, I dropped, I scrambled and I fell. As it got darker I found myself sliding from one dark, barely visible tree to the next, grabbing at each to try and break my fall while simultaneously trying to avoid ripping my face off on the lower branches. It was then that my mother and grandmother's knitwear really came into its own. My balaclava and jumper may not have been comfortable or stylish but they afforded the same level of protection as chain mail.

Shortly before it was totally dark I shot out of the trees at the bottom of the cliff's final slope and dropped on to the road. The knot of fear dissolved from my chest to be replaced by utter exhaustion. I was about six miles from the village which, compared to the normal Long March, was a luxury warm-down for me. When I finally got back to the village, my teacher was very keen to know where I'd been as he was beginning to be concerned that I was lost or injured on the mountain. I managed to convey that I'd had a romantic drink with some fur-clad beauty high in the mountains. Later, in the mirror in my room, I discovered that my face was covered in cuts where my face had been whiplashed by the tree branches I was ploughing through head first. He must have thought I'd had a hell of a date.

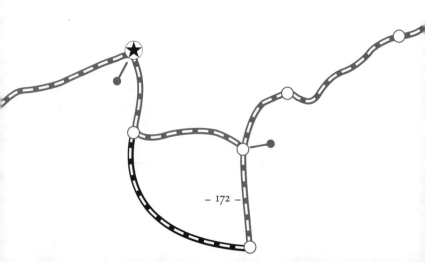

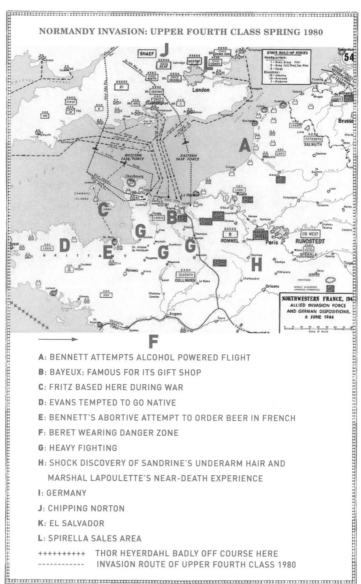

NORMANDY INVASION: UPPER FOURTH CLASS SPRING 1980

A: BENNETT ATTEMPTS ALCOHOL POWERED FLIGHT

B: BAYEUX: FAMOUS FOR ITS GIFT SHOP

C: FRITZ BASED HERE DURING WAR

D: EVANS TEMPTED TO GO NATIVE

E: BENNETT'S ABORTIVE ATTEMPT TO ORDER BEER IN FRENCH

F: BERET WEARING DANGER ZONE

G: HEAVY FIGHTING

H: SHOCK DISCOVERY OF SANDRINE'S UNDERARM HAIR AND
MARSHAL LAPOULETTE'S NEAR-DEATH EXPERIENCE

I: GERMANY

J: CHIPPING NORTON

K: EL SALVADOR

L: SPIRELLA SALES AREA

++++++++++ THOR HEYERDAHL BADLY OFF COURSE HERE
----------- INVASION ROUTE OF UPPER FOURTH CLASS 1980

CHAPTER 17

Normandy, France

I owe France an apology. I know it is the land of high culture, fine living, arts, letters and a thousand cheeses, but I've never really warmed to it. One of the six witticisms* owned, lovingly polished and regularly aired by my Uncle Clive was that France was where Good Americans went when they died. I heard this homily very early on and it just left the unsettling impression that France was full of good but dead Americans.

My main influence growing up was the Second World War and, as the French played very little part in that particular event, they didn't really register with me as a proper nation until very late in the day. On the other hand, a comparatively small country like Norway scored very highly with me because of the heroes of Telemark. For the handful of sheltered and wilfully ignorant people who don't know about the heroes of Telemark, they were a small band of intrepid Norwegians who blew up a German plant making heavy water for

---

* The other five are:
1. Show me the man with his feet firmly on the ground and I'll show you a man who has trouble getting his trousers on.
2. If you fall down and break your leg, don't come running to me.
3. Beauty is only a light switch away.
4. Life is like a box of chocolates. You can't get to the second layer until you've finished the first.
5. Why did the illegitimate chicken cross the road twice? Because it was a double-crossing bastard [my favourite].

the Nazis' nuclear weapons programme and prevented London from being the first city to be nuked. Thank you very much, Norway.

I am also rather ashamed to say that one of the reasons I have never grown to like France is because I have never really got to grips with the language. I understand from those who have mastered it that talking French fluently is akin to the sensation of kissing yourself vigorously with the tongue. I generally deploy a range of excuses for my failure in this department. The first is that my French teacher at school couldn't speak it either. Well, he might have done but we never got the chance to find out. In our first French lesson he informed us that he was a Communist but then immediately went on to say that if ever we spotted his beliefs in any way influencing his teaching then we were to point it out immediately. This, however, was unlikely to happen because, firstly, we didn't know what Communism was and, secondly, he didn't know what teaching was. In order to remedy the first problem and to make sure that we could spot any subliminal Communist messages, he spent most of our French lessons thoroughly inculcating in us the intellectual underpinnings and glorious history of Communism.

Here I need to write briefly about berets. I generally like to think of myself as a tolerant and broadminded kind of person. Difference, diversity, divergences, diasporas and Divali are all OK with me but there is one line I will not cross and that is the line on the other side of which people wear berets. How France can claim to be a capital of fashion while the production, advertisement and wearing of berets has yet to be outlawed is absolutely beyond me. The only thing I find even more baffling is that the beret in various colours has become the much sought after and in some cases legendary symbol of Special Forces round the world. I suppose you have to be exceptionally tough to wear one without flinching, and no one's going to be in a hurry to laugh at you. It won't come as much of a surprise to you that our

Communist French teacher wore a beret, and the very worst kind at that, the one with the little tomato stalk poking out of the top.

The cultural figure who most influenced my understanding of the French was Captain Hornblower. Long before Patrick O'Brian's Captain Aubrey went to sea, C. S Forester's Captain Horatio Hornblower was giving the French Revolutionary fleet an absolute pasting on each and every one of the seven seas. The Hornblower books were always very clear about French warships: they were structurally rotten, their planking like some crumbling old Roquefort, and with filthy, holey sails like a tramp's underpants. In a light breeze you could smell the foetid garlic stench of the French fleet before their sails topped the horizon. Their ships were invariably captained by vicious bullying sexual degenerates and crewed by the dregs of French society emptied from the country's prisons. When engaged in close-order combat, the French, reeking of cheap brandy, fought like girls and generally disgraced themselves. I imagined the French ship of state and all its citizens aboard to be not dissimilar.

Uncle Clive, who was also a bit of a Hornblower buff, had a key fact about the respective merits of the French navy and the Royal Navy. This particular uncle always seemed to enjoy life immensely even though others nearby never seemed to be having quite so much fun. I'm sure his continual state of happiness was due to his possession of a number of key facts that were sure stepping stones in a swampy world.* His key naval fact was that in the heat of the battle

---

* Uncle Clive only had about six phrases and twenty key facts to sustain him across all social occasions, but on the plus side he was immensely strong. Instead of fiddling around with difficult small talk, he much preferred to lift a wheelbarrow above his head using only one hand. This wasn't a little flower border wheelbarrow either, but a large builder's barrow with pneumatic tyre. If caught indoors with a stranger and without a wheelbarrow, he would substitute a number of rapid one-handed press ups, instead of the normal handshake.

at Trafalgar the French managed to run their guns out once every four minutes but the British managed it every minute. Uncle Clive managed to run this statistic out once every four hours, alternating it with another key fact that Napoleon's penis was one inch long fully extended. My view of France, the French and the whole panoply of French culture didn't stray much from that given me by Captain Hornblower and Uncle Clive.

One affable, clean-living French person would have cleared up much of my total misunderstanding of that great country in the same way that Fritz being a decent sort had mitigated the impact Germans had on my imagination, despite them being tiresome throughout both World Wars. Unfortunately, the only French-style person I came into contact with at school was Marshal LaPoulette who was as close to a rotting French warship as it was possible to be. Marshal LaPoulette wasn't even French. He was French Canadian but the impression he gave was of being a living dead, good American who had been interred in France for some time but through some kind of administrative confusion was now making his way home.

I don't make it my business to be uncharitable to people, but Marshal La Poulette gave every impression of having been freshly exhumed. We met when we were fourteen but already he'd battered his way through puberty and looked like a trainee honey monster. His preferred dress, whatever the season, was an acrylic black zip-up puffer jacket, that made him look rounder and huger than he already was. The black acrylic, where it hadn't been punctured by cigarette ash, was stained with every variety of foodstuffs known to French Canadians (including seven hundred types of cheese). His hair, which was a totally nondescript dried-dung colour, always seemed to be caught up in some localised tornado and whirled uncontrollably around his head. From a distance it looked as if a cloud of flies were circling his dung-coloured head. Slightly below his hair was a pair of glasses so thick it seemed as if two minute comedy eyeballs had been stuck on the inside of the lenses. The lenses themselves were covered with a thick layer of dandruff in the manner of an idle Christmas snowscene shaker. When he spoke his French accent was so thick it sounded like he was spreading thick Ardennes pâté on rough French bread. I have to say I quite liked him.

There was one thing about Marshal LaPoulette I didn't like and that was, contrary to all appearances, he was good at football. You can imagine how traumatic it was in the playground for Marshal LaPoulette to be picked before me. In a way, that, too, was like France. I don't just mean that they were annoyingly good at football but that even though they're clearly a decaying Third World nation they can still come up with things of staggering vision and beauty like the Millau Viaduct across the Tarn Valley in southern France. This has to be one of the wonders of the modern world. There have been bigger and more expensive buildings but none quite so beautiful. Civil engineering is important to me. I like big things imaginatively conceived and beautifully built. The Fatted Calf shares this appreciation, hence the magnificent Reichstag Steps in our garden, built against the odds with slave labour. I once drove down a new piece of motorway and said I thought it was a lovely piece of engineering. I asked my girlfriend whether it did anything for her. She said it got her to the shops quicker. We parted shortly afterwards.

When I reached fifteen, Uncle Clive, Marshal LaPoulette and Captain Hornblower had all but completed my personal understanding of France and the French. At school, fifteen was considered the age when boys were sufficiently versed in the catalogue of French military humiliations at the hands of the British to be let loose on the country itself without danger of being impressed by their creamy sauces, decadent women and other nonsense. All that was needed was for us to visit the country itself and this, happily, was arranged by our Communist leader as a school trip to Normandy, ancestral home of William the Bastard, and to Brittany, ancestral home of Britney Spears.

Quite by chance fifteen also happened to be the age when we teenagers were completely preoccupied with the hormonal *coup d'état* taking place in our spotty little bodies. Thus, when our little school party approached the magnificent, history-soaked coastline of France, we quite naturally found ourselves in the bowels of the ferry crowded round the condom machine. The instructions in French probably said insert coin and pull lever hard. As you can imagine our limited grasp of French translated this into an act of typical Gallic depravity.

At that stage of our sexual development, ownership of a condom was just about the naughtiest thing possible and very nearly approximate to full-blown sexual intercourse. We were all enormously impressed, therefore, when Bennett, the leading figure in our party, actually bought a packet. The one thing we were all looking forward to on our trip was French women, who in the depravity department knocked their English counterparts into a cocked *chapeau*. (I think I should mention that I went to an all-boys' school so my exposure to women up to this point had been limited to Boadicea). Unfortunately, apart from Bennett's disastrous attempt to order a beer from a waitress in St Malo, our exposure to French women was limited to our tour guide, Sandrine. Superficially, Sandrine didn't exhibit any signs of wanton depravity, but we were all prepared to give her the benefit of the doubt.

Before departure we had been taught that Brittany was once a separate country colonised by the Welsh and Cornish. We knew that Evans was Welsh so we all kept a weather eye out for him in case he disappeared off into the hills singing 'Bread of Heaven'. I felt that a history of colonisation by the Cornish would mean a proliferation of small ice-cream shops. The fact that there weren't any we put down to the heavy fighting in the area during the war. 'Heavy fighting' was the big phrase of the trip and anything we saw that wasn't spanking new we identified as a tragic reminder of the 'heavy fighting' in the vicinity.

The one cloud over the trip was in the shape of William the Conqueror who, by dint of sneaking up on us from behind while King Harold was dealing with a simultaneous Viking attack, had successfully invaded. We consoled ourselves with the fact that William the Conqueror was in no way French but was a Viking himself (Norse Man, hence Normandy). Steeled with this knowledge we went to see the Bayeux Tapestry. This was singular for the fact that, although immensely long and of obvious historical importance, it was completely devoid of naked women. There was, however, a gift shop where the French balance of payments received a welcome boost by the purchase of twenty-four identical postcards showing King Harold with an arrow pointing to the location of his fatal injury.

Mont St Michel, a cheap imitation of St Michael's Mount, was more evidence that Brittany was colonised by the Cornish. Having marched straight to the top and back in five minutes, we had ample time to discuss all the possible angles of the Sandrine situation and the latent wanton depravity therein. Evans had been caught trying to speak French with her. We didn't know whether to be more worried about him going native or getting sexually out of his depth. Later, the Gothic splendour of Chartres cathedral was almost completely eclipsed by the discovery that Sandrine had underarm hair. We all associated untamed body hair on women with extreme moral turpitude (our mothers mostly lived their lives according to the moral guidance of *Good Housekeeping* magazine, where an unmown armpit in the house would be as unlikely as an unmown lawn outside the house).

It was somewhere near Chartres that Marshal LaPoulette had his near-death experience. Just down the road from the hotel where we were staying that night, there was a little nightclub. Because we were extremely mature we were permitted to go there as long as we were back by eleven. This was, of course, only about half an hour after the management had finished unlocking the front door but it did nevertheless give us English boys a chance to jump around like idiots for a while. Marshal LaPoulette was one of the first on the dance floor and, even though he looked like a mobile compost heap, he danced like Michael Flatley. He was an enigma, that Marshal LaPoulette: good at football, great at dancing and probably stupendous in bed but visually straight out of the back of a garbage truck.

After a while the dance floor began to get crowded and it was at this stage that Marshal LaPoulette suddenly fell to the ground and remained there, motionless. The organisers muscled their way on to the dance floor and, believing him to have fainted, and with no regard for their personal safety, picked him up and carried him off the floor. As he was being carried shoulder-high from the dance floor by four large Frenchmen, Marshal suddenly shot out his arms and screamed 'Yahooooo!!' at the top of his lungs. He was dropped like a hot brick and then rolled around the floor laughing like a hyena, seemingly oblivious to the fact that he was about to have his head kicked in by the disgusted Frenchmen. I can only think the reason they didn't do this was that no sensible person would want to get their shoe caught in the mess on top of Marshal LaPoulette's head.

A few days later at Le Touquet, the scene of heavy fighting, we were told that it was the favourite haunt of the upper-crust Edwardian English. Burrows thought he felt a lingering sense of *fin-de-siècle ennui*. After we removed his glasses and zipped his head into his Adidas bag, the feeling soon passed and we heard no more about it, or anything else from Burrows. As we were near the Loire region, and as a special concession to our obvious maturity, we were allowed a glass of wine with our meal. Bennett, who was big for his age, and also rather stupid for any age, drank three glasses and later fell out of his bedroom window. Of course his parents had to be informed, but as information never helped Bennett much, I'm sure it didn't help his parents either. On our departure Sandrine kissed Bennett on both cheeks. There was some pretty heated discussion on the ferry as to whether this constituted a French kiss as properly defined. Having backed Bennett wholeheartedly in this discussion I was given one of his condoms to take home with me, which sort of made the trip worthwhile. I believe I still have it somewhere.

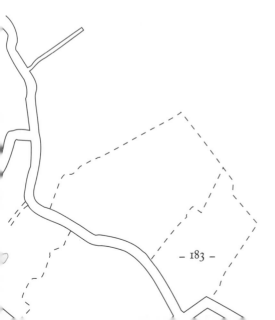

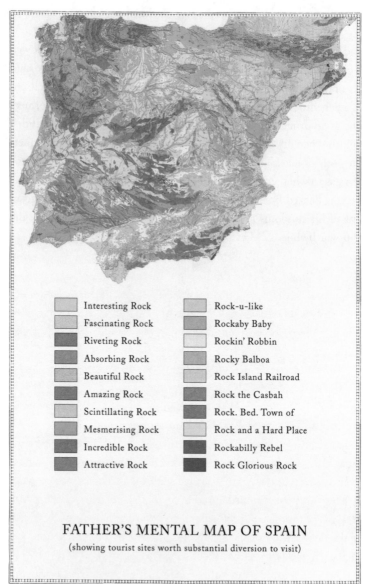

Interesting Rock   Rock-u-like

Fascinating Rock   Rockaby Baby

Riveting Rock   Rockin' Robbin

Absorbing Rock   Rocky Balboa

Beautiful Rock   Rock Island Railroad

Amazing Rock   Rock the Casbah

Scintillating Rock   Rock. Bed. Town of

Mesmerising Rock   Rock and a Hard Place

Incredible Rock   Rockabilly Rebel

Attractive Rock   Rock Glorious Rock

FATHER'S MENTAL MAP OF SPAIN

(showing tourist sites worth substantial diversion to visit)

CHAPTER 18

Minorca, Spain

Finally, shortly before the Fatted Calf left home for college, our family went on its first package holiday. Every part of it was a new experience for us. We flew on a plane from Luton airport with lots of other people to a holiday somewhere hot, Minorca, to accommodation not owned by an academic colleague. To us children Minorca finally meant sunshine; it meant sand, it meant sea, it meant swimming pools, it meant sun-tan cream, buffet lunches, Coca-Cola with straws, other children to play with/snog, sightseeing, organised evening entertainment and all the fun you could eat. It was going to be one hell of a holiday.

To my father Minorca meant limestone. As we discovered almost immediately on arrival it has some of the most spectacular outcrops anywhere in the world. Why, you could even see some incorporated into the 'Welcome to Minorca' sign at the airport. Almost unbelievably the next day found us on a Long March into the interior in search of spectacular limestone outcrops. There was a stark contrast in mood between my father armed with his geological survey map striding into the parched interior of the island beyond the white breakers of timeshare apartments that washed the shore, and the rest of the family who looked longingly back at Los Pinguinos hotel on the beach with its ice-cold Coca-Cola and straws.

One of the things my father found hardest to deal with was the fact that his passion for limestone wasn't shared by the general populace. He found this particularly baffling in Minorca, where the locals were, after all, sitting on some of the world's finest. On our Long March (I remember there were Lincoln biscuits on this march, too, so it must have been pre-arranged even before we left home) my father insisted on accosting various bemused locals and engaging them in lively discussion about the wonderful geology they were blessed with. These discussions were lively for reasons other than shared passion for sedimentary deposits in the Jurassic age. The first reason was that my father's Spanish, while fluent, was also coloured by his long exposure to Central American peasantry and his love for their ingenious ways of swearing. Spanish Spanish speakers find Central American Spanish quaint at the best of times so they must have found an English academic swearing at them in El Salvadorian peasant argot particularly trying, especially as he seemed to be talking complete rubbish about rocks.

It must have been after several of these lively discussions, when we were about twenty miles into the parched interior of Minorca and our single bottle of squash was completely empty, that my mother put a stop to the Long Marches for good. We returned to the hotel in a car that wasn't a taxi. On rare moments of industrial-strength incompetence by my father, my mother takes charge. The effect of this is like constituting COBRA, the Government's emergency and disasters committee. Things happen very quickly at a very high level through shadowy and powerful agencies that it's best not to ask too much about. I'm sure she commandeered a passing vehicle and, with Spanish that was more in keeping with Queen Isabella of Aragon's, ordered the innocent motorist to return us immediately to Los Pinguinos without further let or hindrance. I hope the poor driver knew there was a hotel called Los Pinguinos otherwise he would have had a formidable English woman demanding that he 'return her to the penguins', which he may well have assumed by her demeanour was her natural habitat.

For the rest of the week we had to fall back on the pleasures of the resort which mainly involved doing nothing under the sun while waiting to eat. In the absence of a bedrock of limestone and Long Marches, the holiday began to disintegrate. The pattern for all our previous holidays, excursions and visits had been to go somewhere and learn something, as opposed to the package holiday which assumed you would go nowhere and do nothing. After about three days of this my father had the equivalent of a mental breakdown and took us to the local capital, Port Mahon, which had no beaches, no crazy golf, no swimming pools and, even more importantly, was a port used by the Royal Navy in the eighteenth century.

In the absence of limestone and peasants my father is equally happy to talk about the Royal Navy, think about the Royal Navy or visit anything with any kind of connection to the Royal Navy. In my youth I think I spent more time on HMS *Belfast* than some of the men who served on her. HMS *Belfast* is a big, old British Second World War Town-Class light cruiser, moored on permanent display on the Thames in central London. I liked it because it was a cruiser and had guns; my father liked it because it gave him an opportunity every five yards to remind us of his glorious naval career; and the Fatted Calf liked it because there was a small shop that sold HMS *Belfast* rock which could be sold on at a healthy mark-up at school.

One of the things our father used to tell us with Swiss regularity was that not so long ago boys of our age were sent to sea as midshipmen. The Fatted Calf and I both took this as a very clear signal that round the next bulkhead there would be a Midshipman Recruitment Desk and that we would both be signed up for a fifteen-year apprenticeship. My father was also very keen to show us the bridge where he would have worked as a signalman, but I remember this being closed, which we very quickly put down to his presence on the ship and the possibility of some signalling catastrophe ensuing.

An unfortunate legacy of my father's time in the Royal Navy was his predilection for boat trips. You have to remember that I was sick when my grandmother's Triumph Herald took a roundabout, so being afloat was a deliberate incitement to vomit as far as I was concerned. Nevertheless, we found ourselves boarding various vessels with depressing regularity. In London once, after the routine inspection of HMS *Belfast*, my father took it into his head that we should take a boat upriver to Hampton Court Palace. The tour guides, especially the ones who own the tourist boats, give the impression that Hampton Court is just the other side of London but what they neglect to mention is that the Thames is exceptionally wiggly through London. As everybody knows, if you remove your intestines and straighten them out they're actually ninety miles long. It's the same with the Thames through London: if you straighten it out and work out how far you've actually travelled, it's the equivalent of taking the ferry from Felixstowe to Spitsbergen in Norway. That's why even good travellers feel like death by the time they arrive at Hampton Court.

Many years later I lived in Hampton Wick, a small village on the Thames just downstream from Hampton Court. When my father had his sixtieth birthday, he thought, in recognition of his advancing years, that he would walk the entire length of the Thames from Oxford to Hampton Wick. Following the principle of the straightened intestine, this particular Long March was not far short of Chairman Mao's original Long March across the length and breadth of China. As a surprise for my father when he arrived at my house, I'd arranged for a little steamer to take him and his brothers and sisters, my uncles and aunts, on a little river cruise with a picnic.

The steamer was a veteran of Dunkirk, one of the little ships that had volunteered to cross the Channel and pluck our lads from the beaches of Belgium. As Dunkirk was one of the Royal Navy's finest hours, this was a particularly exciting development for my father.

Sadly, this particular little ship managed to survive the Luftwaffe's attentions at Dunkirk but not the Browning family's. We lost power mid-stream, the engine blew up and, after drifting helplessly for a while, another steamboat arrived to take us off. Fortunately my father's naval training kicked in and we were able to transfer a good proportion of the aunts safely from one ship to the other. For once in his life, Uncle Clive's legendary ability to lift a wheelbarrow over his head came in useful as he propelled aunt after aunt from our little ship over the transom of the much bigger ship in an action very similar to shot-putting.

Going back to limestone – and to be absolutely fair to my father, rocks aren't necessarily a bad basis for a holiday. It's all a question of expectations. At the beginning of the very same summer as our package holiday, I went on a school geography field trip with the sole intention of looking at rocks. We spent a week looking closely at the ground which is what most teenagers do anyway, but this time we had to focus on it. A whole day was spent measuring the size of the pebbles on Chesil Beach in Dorset. This is a particularly long beach richly endowed with pebbles of all sizes. There was a learning to be had from that but the only one I remember now is that there are a hell of a lot of pebbles on Chesil Beach. All geography field trips in the South of England go to Chesil Beach, which means there is a frightening amount of measuring, grading and general interference with the pebbles. Indeed, there's a Ph.D. waiting to be written on the impact of geography field trips on Chesil Beach.

Mr Renshaw, our geography teacher on that trip, was an exceptional man. I have a very poor memory but I do remember him and even something he taught me. Teachers often worry about 'the mind of the examiner'. Mr Renshaw didn't have to worry about it because he had it. He taught like a perfect exam answer. It was wondrous to behold. After each lesson you felt you'd come out of an exam paper in which you'd done exceptionally well. No one ever missed his classes as this was tantamount to failing the exam. Everything he taught was an answer and even the way he taught us to think was guaranteed to tick boxes. His motto was 'Fact, Reason, Example'. For example: beaches have pebbles. They have been eroded by the sea. Chesil Beach has lots of them. Wallop, that's GSCE Geography in the bag. To be honest, I struggled to come up with that particular example and I'm beginning to suspect that Fact Reason Example doesn't work anywhere in life outside Physical Geography. I think it works better the other way round: Example, Reason, Fact. My wife has seventy pairs of shoes; she likes shoes; we can't afford a holiday.

I nearly had a career in rock. This may surprise you as nothing you've read so far has been pointing in the direction of heavy metal. The rocks I'm referring to are actually stones, dry stones to be more precise. Mr Lancaster, who I mentioned earlier as being the dominant feature of Botley, was a legendary figure. He was one of the shop stewards at British Leyland's Cowley plant in east Oxford, where they made design classics such as the Austin Allegro and Morris Marina. Mr Lancaster was an absolutely lovely man and a pillar of the community, while simultaneously and single-handedly sabotaging car production at the Cowley plant throughout the dark days of the late seventies. Mr Lancaster could find ways of going on strike when he was already on strike. In his way he was a genius; he knew more about labour law, industrial relations, manufacturing and personnel policies than anyone else on the site and used his vast experience to make sure absolutely no work was done and the very minimum of cars were produced at the very lowest quality. He was an incredible leader and could make thousands of men down their tools as one without a murmur of complaint.

The other striking thing about Mr Lancaster (if you'll forgive the pun) was that he was enormously accident-prone. Working on a car production line is clearly a fairly dangerous job, but Mr Lancaster must have had a particularly poor internal sense of Heath and Safety. He was forever getting bits of himself caught in the line. I know he lost several fingers that way and my abiding memory is of him either walking round the estate on crutches or with his arm in a sling, or both. Often it was difficult to tell whether he was on sick leave or industrial action. I wouldn't be at all surprised if he organised a strike for his own personal protection so that he didn't lose another vital part of his body while working. Despite the fact that he did very little of it, he clearly loved his job and his workmates. His missing parts he wore as badges of honour in his struggle on the front line of British industry and his continuous efforts to bring it to a complete standstill.[*]

Only two things hindered my total hero worship of Mr Lancaster. The first was that he used to come swimming with the Cub Scouts as one of the parent helpers. He was around more than most parents because his place of work was almost permanently chained up and guarded by pickets installed by himself. Mr Lancaster seemed to like swimming a lot and was very helpful to us learners. I noticed after a time that, when Mr Lancaster swam himself, he was very high in the water, almost in the manner of a hovercraft. As his favourite stroke was breaststroke, I assumed that he gained incredible lift from powerful arms that were used to lifting heavy engine blocks and putting them down smartly as another strike was called. Then, one day, Trevor Adams lent me his swimming goggles and for the first time I could see clearly underwater. I wish Trevor had never lent me those goggles because, just before I submerged, I saw Mr Lancaster

[*] BMW make Minis in the Cowley plant now and it's a model of efficiency. Something makes me feel that they wouldn't tolerate characters like Mr Lancaster, which is sad in its way. Still, there must be some corner of that huge factory that still has a little part of Mr Lancaster, probably one of his fingers.

powering past in his hover breaststroke style and, just after I submerged, I saw Mr Lancaster's right leg hopping gamely along the bottom, like an ostrich fording a river rather deeper than it had anticipated.

The other inexplicable and curiously disappointing side to Mr Lancaster was his father, an elderly man living with his family behind a door in their sitting room. Whenever I went to visit their house, Grandpa Lancaster was in the same position behind the door and whenever the door opened it banged into Grandpa Lancaster's chair. Even though I was very young I remember thinking that this was no way to treat the senior member of the family (or a chair, for that matter). If I'd been Grandpa Lancaster I'd have come out on strike. Knowing that what comes round inevitably goes round, I've got a terrible feeling that somewhere, probably in Botley, the great Mr Lancaster himself is probably now wedged behind a door in his son's house, his chair regularly bumped by all and sundry.

None of this in any way detracts from the superb piece of career advice Mr Lancaster was kind enough to give me. He'd singled me out at a very early age as being the kind of boy who would grow up practically unemployable (wise man). One day after Cub swimming he sat down beside me and told me that if I wanted a really good job for life then I should be thinking very seriously about either going into thatching or drystone walling. There is certainly a market for both of those things even now but not, I fear, for the kind of stone wall or thatched roof I would have been capable of throwing together. Looking back on it, he may also have been thinking that thatching, drystone walling and other traditional crafts were going to be all that was left after he and his comrades had finished laying waste to British industry. Either way, it was well-meant advice and I sincerely hope he's fit and well and has avoided being wedged behind the door.

From my preparatory homework for the Minorca trip I remember reading that there was so much limestone on the land that much of the history of that island was occupied with the back-breaking task of clearing it from the fields. When they cleared a field they used the limestone to make long stone walls around the cleared plot. Had I listened to Mr Lancaster I could now be on a permanent working holiday over there, tidying up and rebuilding the walls and occasionally breaking for a Coca-Cola with a straw.

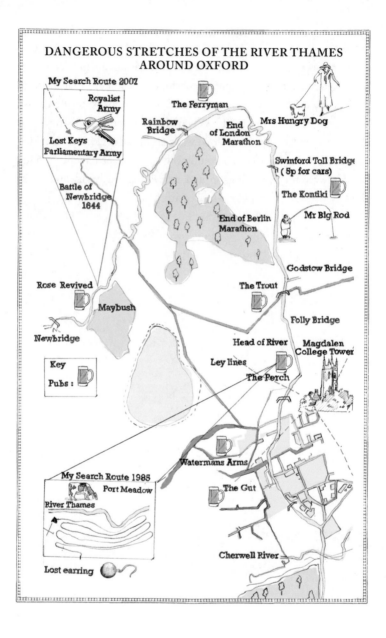

DANGEROUS STRETCHES OF THE RIVER THAMES AROUND OXFORD

My Search Route 2007

Royalist Army

Lost Keys
Parliamentary Army

The Ferryman

Rainbow Bridge

End of London Marathon

Mrs Hungry Dog

Swinford Toll Bridge (5p for cars)

The Kontiki

Battle of Newbridge 1644

End of Berlin Marathon

Mr Big Rod

Godstow Bridge

Rose Revived

Maybush

The Trout

Folly Bridge

Newbridge

Head of River

Magdalen College Tower

Key

Ley lines

Pubs :

The Perch

My Search Route 1985

Port Meadow

Watermans Arms

River Thames

The Gut

Cherwell River

Lost earring

CHAPTER 19

The River Thames, Oxfordshire

Sacrilege is not something you come across on a daily basis, especially not in the Botley area, which is why the event I am about to relate was so profoundly disturbing. Books were treated with a great deal of respect in our house. When someone was reading a book a hushed silence was encouraged from everybody within a quarter-mile radius. Physically, books were also accorded respect; laying a book face down instead of using a bookmark was considered in the same light as the pile-driver move in wrestling, i.e. a deliberate attempt to damage the spine and therefore totally *verboten*.

Books also acquired more attention in our household by default. This was because for long periods we didn't have a television. In my early years we did have a television because I clearly remember pictures of the Vietnam War which the Americans were in the process of losing rather badly thanks, no doubt, to Dr Kissinger and his bright ideas (or possibly could have won if only they'd listened to the great Dr Kissinger). More importantly to me but perhaps less important geopolitically, I also remember *Tales of the River Bank*, starring Hammy Hamster, which would have been a charming children's programme had it not also featured a guinea pig piloting a small plane, which was absolutely preposterous.

Our television was taken away just before the Fatted Calf's eleventh birthday and his graduating to Big School. My father thought a television would interfere with our advanced studies and, as he hadn't had one when he was our age, there was no reason why we should have one. We pointed out that television had yet to be invented when he was our age but he regarded this as 'insolence', a catch-all description of any behaviour that didn't tend towards silence and absence on our part. I could tell you the exact time and date our television was taken away if I bothered to look it up. It was the FA Cup Final of 1974.

We don't really follow football in our house principally because none of us can play it as it involves extremely complex coordination between hand and eye. Or maybe foot and eye (perhaps that's where I've been going wrong). I remember the exact moment I stopped being interested in football: I was about seven, wandering aimlessly around the football pitch while the rest of my class thundered about chasing the ball. By some fluke the ball detached itself from the chasing herd and travelled straight over to the empty quarter of the football pitch were I was minding my own business. Without exaggerating, I probably had about an hour to decide what to do with the football that was heading directly towards me, followed by the baying mob. After a moment's reflection I decided to panic, flailing wildly at the ball which rolled politely past me totally unmolested. I then completed this impressive manoeuvre by falling over. The teacher bellowed across the length of the pitch, 'Browning, you are a wet lettuce', and I mentally gave up on football with the consequence that England have since failed to win a major championship.[*]

---

[*] As children we were never taken to see football matches on the fairly reasonable grounds that they were a total and utter waste of time. I have only ever seen two football matches in my entire life. Once I was taken by an Old Etonian to see Arsenal play Norwich at Highbury. It was very cold and the score was 0–0 and my lasting impression was that an Eton education had clearly done him no favours. I also went to see England play Italy at Wembley. The score was 0–0. As they couldn't be bothered to score a goal for an hour and a half despite me paying an emperor's ransom for a ticket, I transferred my affections to banger racing, where the score is never 0–0.

Nevertheless, if I'm asked what team I support I generally say West Ham because if you say, 'I don't support any football team because football is played, watched and financed by complete idiots', you tend to lose friends and be subjected to the pile-driving manoeuvre by big football fans. My father supported West Ham, as did his father, because that's the nearest proper team to Loughton unless you include Leyton Orient, which no one outside Leyton does. West Ham were in the FA Cup Final of 1974 and I was watching it on our television. The men came to collect our rental television at half time during the FA Cup. Looking back, they must have been the only rental television men in the country who were working on a Saturday and had no interest in football. Needless to say West Ham won that final 2–0 in the second half but by that time I was staring at an empty area in the living room across from our sofa.

The two consequences of the television leaving were that we were thrown back on books and maps to keep ourselves amused. Also I was suddenly cut off from popular culture and found conversation in the playground much more difficult unless it took on a retro slant with a discussion of the glory days of Hammy Hamster. The point of me telling you all this is that books were important in our house and maps were esteemed even above books. That is why it was such a shock for me one day to open an Ordnance Survey map of the Oxford area and to find that my father had drawn long straight lines all over it in bright red biro.

It turned out that my father had become interested in ley lines, which are a kind of invisible geography for the spiritually minded. Believers in ley lines will tell you that ancient churches and Neolithic monuments such as barrows and henges are aligned along lines of energy that our forefathers knew about but that somehow we have forgotten in our worship of the scientific method. My father took to ley lines in a big way. He had all the fervour of a convert and began to draw long lines across all sorts of valuable maps, which in terms of supplanting one object of worship with another was akin to the Taliban blowing up statues of the Buddha. With a bit of creative draughtsmanship he managed to get all the sacred sites in Oxfordshire lined up and he found that the lines intersected on the tower of Magdalen College in Oxford – interestingly, still the site of pagan rituals on May Morning.

One of the worrying things about parents is that they often provide reliable blueprints of what you're going to be like when you get to their age. This may explain why I found myself, thirty years later, attending a course on dowsing, which is the study of lines of energy and how to detect them using nothing more than a coat hanger, a pendulum or a couple of bent twigs. I thought dowsing was something that only eccentrics and Professor Calculus in the Tintin books did until I saw a very normal looking workman park up in a Water Utility van, jump out with two rods and proceed to find the underground pipe he was looking for. When I asked him about it, he said they all dowse for pipes as it's a lot faster and more reliable than using the ground X-raying machines provided for them at great expense.

When something is dismissed outright by the scientific establishment, it's a sure sign that it's something worth studying. The course I went on studied energy fields, auras, ley lines and all sorts of other fascinating stuff. We went out and about doing all manner of practical experiments dowsing for underground water and pipes and energy fields. Everyone was having a high old time, with their rods crossing and recrossing as they passed over pipes and cavities and anything else they chose to look for. Some people were using a pendulum instead of rods and these were swinging this way and that like the tongue of an excitable dog. It pains me to tell you that I was the only person on the course who couldn't do any of it. My two metal rods refused to cross, to move or even to twitch. I could walk over a thirty-inch copper pipe with metal stanchions and my rods would be as parallel, motionless and resolutely erect as the tines of a carving fork.

By far the most adept individual on the course was a nutty little man of about fifty who'd been servicing boilers all his life. He was so attuned to lines of energy and the presence of water that his rods would spring into life every time I walked past him with a cup of tea. He'd done other divining/dowsing courses before and he told me that he no longer bothered to look closely at the boilers he was servicing. He knew all the components of a boiler and would simply run his pendulum up and down outside the casing to see what bit wasn't working. He'd then open the boiler, replace the part and spend the next half an hour sucking his teeth and pretending to adjust various bits while he dreamt up a figure to put on his invoice. The rest of the group had more or less mastered their dowsing techniques by the time the course finished. Except me, of course, who still had chronic frozen-rod syndrome.

Eventually the course leader had to take me aside for some special remedial lessons such as seeing if I could detect a large bucket of water sitting on the desk in front of me. As I struggled with this and later, even simpler, more humiliating tasks such as spotting the swimming pool in front of me, he gave me some advice. He said my problem was that I was thinking about it too hard and that I just needed to relax. Apart from football, the two things which I am worst at are relaxing and not thinking about things. All he had to do was to tell me to simultaneously keep a ball in the air using just my feet and head and the impossibility of me ever dowsing would have been complete. I finished the course having understood a great deal but without being able actually to do anything. They very kindly gave me a certificate but inked out most of the things that it certified. I don't think they signed it either.

None of this stopped me giving the impression, once I'd returned home, that I could now find needles in haystacks, Holy Grails and anything else that was secret, hidden or lost. Within days I'd managed to prove this beyond my wildest imaginings and, as you've probably already noticed, my imaginings can be pretty damn wild at times. I happened to be attending a recreation of a Civil War battle done by a remarkable body of people called the Sealed Knot. These are men and women from all walks of life who feel a pressing need to bring to life the English Civil War and to re-engage in some of the battles involved. An essential part of this is to dress in exact replicas of the clothing worn on the day and to fight with only the weapons available on the day.*

A close friend of mine was involved in this bizarre cross-dressing rambler cult – I beg your pardon, valuable historical re-enactment society – and was helping to organise a battle at Newbridge on the upper reaches of the River Thames, the site of an important Civil

---

*The American equivalent of this is called anachronistic simulation. It's a bad name and results in bad practices, such as recreating medieval battles in Arizona. Quite by chance I happened to attend a re-enactment of Custer's Last Stand at Little Big Horn in Montana when I was touring America trying to find myself. As we all know, and the dowsing course confirmed, I can't find things for toffee, me included, so in some ways it was a wasted trip. However, on my way back through Montana I took a classic Browning short cut and found myself extremely lost in the middle of the Crow Indian Reservation. I had to sleep in my car outside their trading post (OK shop but trading post sounds better). I was befriended by a young Native American boy whose name turned out to be Guildford Seize the Ground. You can't make something like that up. Well you can, but it wouldn't be true. When he heard me speak he assumed I was some kind of alien life form and after I'd explained I was English this seemed to confirm it for him. I told him that there was a city in England called Guildford and that we only showed western films where the Indians won quite handsomely and the palefaces ended up in reservations. All in all, I think he benefited from our conversation and seemed sad to watch me drive away the following morning.

War skirmish in 1644. On the day before the battle, my friend lost her car keys in the main car park, which at that time was a massive open and empty field stretching down to the river.

When I arrived at the car park she and some of her other friends were wandering up and down with their eyes on the ground, like anxious chickens looking for grain. I have to stress at this point that it was a *very* large field, with quite long grass, which made finding a small bunch of keys virtually impossible. In the interests of total fairness, I should also stress that at this time I had her keys in my pocket for a complicated set of reasons that are best forgotten and had fortunately been completely forgotten by all the people involved in the search. By a happy coincidence I also happened to have a pendulum in my other pocket which I'd purchased at the end of the dowsing course in the anticipation that one day I would accidentally stop thinking, relax and strike oil shortly afterwards.

I told the assembled searchers that I would attempt to find the keys using my pendulum and, by implication, the phenomenal extra-sensory powers with which I was abundantly blessed. Not wanting to get their hopes up, I modestly reminded them that dowsing was an inexact science and that it wasn't one I was very good at anyway. I then proceeded to undertake a complex series of manoeuvres with the pendulum as I walked up and down the field, keeping the inert little sod moving with surreptitious flicks of my wrist. Once I had done about five minutes of this, allowing me to get to the middle of the field and away from the continuous barrage of derision, sarcasm and hooting laughter coming from my so-called close friend and her weird cross-dressing friends, I bent down low, fished the keys out of my pocket and announced to the astounded searchers that I had located them.

They were pretty impressed, to put it mildly, and I've since acquired a bit of a reputation as being a minor Gandalf-like figure in the neighbourhood. Wisely, I haven't volunteered to find anything else, as my extraordinary extra-sensory powers take time to recover from a successful outing such as the miraculous Newbridge find. I've never told anyone that I cheated like a cheating cheater but, to my way of looking at it, after the totally unappreciated earring search for my North Oxford girlfriend I was owed credit for one big search and one big find in the cosmic scheme of things and I now had it. And that, of course, is how lines of energy work in the cosmos. Take it from me, I have the certificate.

The River Thames has always seemed to flow through my life. In Botley I could see it from my bedroom window and occasionally it would flood the entire valley below. Later on, when I moved to London, I lived in Fulham, roughly opposite the start of the Boat Race and then upstream near Teddington, where the River Thames first becomes tidal. Because I've always been a bit of a runner I've spent an inordinate amount of time running along the banks and towpaths of the River Thames.

There are three communities of people on the average towpath: anglers, dog walkers and runners. The sole purpose of the anglers and dog walkers is to cause injury or death to runners. Dogs are always well behaved until runners come into view. To a dog a runner in shorts must look like Meals on Wheels and they get correspondingly excited. This excitement always involves pulling their lead tight across your path with the result that you trip over and go head first into the river. The worst part of this is to emerge from the river with seaweed on your head to hear the dog owner say, 'Naughty boy!' This is one rung above the other thing that dog owners say when their hound is swinging from your gonads by its teeth: 'Don't worry, he won't hurt you!'

I know I'm on dangerous ground here as saying anything bad

about dogs in Britain goes down as well as admitting to membership of the Waffen SS but that particular gonad lunch incident put paid to my London Marathon training of six months. The following year a fisherman put paid to my Berlin Marathon. Fishermen in this country basically take the contents of their living room down to the riverbank and sit there in total comfort without the company of their wives or, as far as I can make out, fish. They then spend literally hours watching for the slightest movement on their portable DVD player. Very occasionally they will reach forward to open their monster thermos flasks but otherwise they are completely motionless.

I was running rather smoothly down a towpath near Oxford when one of these anglers decided to make a once in a lifetime sudden movement. Fishermen seem to have two long things when they fish: one is clearly a rod but the other seems to be a long pole of indeterminate function that sits alongside them. This is generally laid right across the towpath to get in everyone's way but is not a problem to hurdle. In this instance the fisherman decided to move the rod when my leading leg was over it but not the following one. Needless to say, I didn't clear that particular hurdle and ended up in the bulrushes like a returning Moses. The angler was horribly upset as I'd clearly woken his maggots or something. I wasn't really concerned about his feelings because I ended up with an ankle the size of a puffer fish and that was the closest I got to Berlin.

SHOOTING STARS OVER CAMP McDOUGAL
PORT JERVIS, NEW YORK STATE, USA
AND SOME OF THE WISHES I MADE

1. Immediate unconditional surrender of Elisabeth de Wilde
2. World peace
3. End of world hunger
4. West Ham to win FA Cup
5. Cash back from Fatted Calf
6. Competence in Michael Jackson Moonwalk
7. Cure for cancer
8. Improvement in hand-eye co-ordination
9. Jensen Interceptor for very little money
10. Long life for extended family
11. New Walkman
12. Unified theory of Relativity and Quantum Mechanics
13. Meeting with Debbie Harry
14. Rapprochement with Argentina
15. End to travel sickness
16. Deborah Wills my first girlfriend to apologise for dumping me for Mark Lawrence

CHAPTER 20

Port Jervis, New York State, USA

Two things brought me back to America: shaving cream and crispy bacon. The Fatted Calf on a trip to America the previous year had brought back some mentholated shaving cream. Shaving with it made your face feel as if it had been born again. It gave you unlimited determination to create a brighter, better world. It made you feel strong, attractive, vigorous, smooth but tough, mature but young at heart. More than anything else, it drew women inexorably towards you as if you were industrial-strength flypaper. At least this was how the Fatted Calf sold it to me when he got back from his trip. I bought a can from him because to refuse would be to give him an excuse to use his high-pressure sales techniques, which included anything from blackmail to physical incarceration. In truth it was excellent shaving cream and I thought the saving I would make buying it myself in America rather than from the Fatted Calf at home would probably pay for the flight over.

Bacon, so I understand it, is the big breaker for vegetarians. It's the torture test for those who have sworn not to eat meat in the same way that pineapple is the torture test for people who, like me, have sworn never to touch fruit and vegetables. The bacon in this country can be quite unappetising, especially if it hasn't been cooked properly. American bacon is, on the other hand, a life-changing experience and surely one of the reasons why there are so few vegetarians over there. Eggs, the demon lover of bacon, are also better stateside. By some incredibly advanced technical process, probably a spin-off from the space programme, they have worked out how to fry eggs without three pounds of lard being involved. Two eggs sunny side up are a couple of incredibly effective anti-depressants, and when they sidle up against crispy bacon, hash browns and this excellent kind of toast that absorbs egg yolk, you have a breakfast which can set you up for anything. And if you've shaved with menthol before, well, you begin to get an idea where that incredible American exuberance comes from.

The town of Port Jervis in upstate New York could not be classified as exuberant by any stretch of the imagination. It's an inland port which, in itself, is a bit of a downer. Ports are never very glamorous but at least they have the sea. Inland ports have neither glamour nor sea. Port Jervis is a port on the Delaware–Hudson Canal. It's also the end of the line, literally, as it's the final stop seventy-five miles out for commuter trains to and from New York City.

Geographically, Port Jervis does have one small point of interest because it is slap-bang in the centre of the 'Tri-State Area'. This rather unappealing name, suggesting as it does somewhere unpleasantly close to the 'Pro-State Area', means that Port Jervis is close to where the boundaries of New York, New Jersey and Pennsylvania meet. There's even a little plaque there to show exactly where the boundaries are and if you assume an unnatural position you can actually have one part of your body in three different states at once.*

---

* Many years later I found myself at Four Corners Point, where Utah, Arizona, Colorado and New Mexico meet. There is another big plaque on the ground and the American I was with absolutely insisted we put one limb in each of the four states. Another fifty or so tourists were intent on doing exactly the same thing so we had to queue for about half an hour while one often very large tourist after another executed a bizarre ritual that looked like elephants coming to a waterhole to die. I can assure you that this doesn't happen very often at the tri-county point between Oxfordshire, Wiltshire and Gloucestershire.

One thing I envy the continentals is their land borders. When you're British you generally have to fly somewhere to get anywhere so the only border you ever see is the little white line six feet in front of passport control at the airport. Crossing from Wales to England over the Severn Bridge is a bit like flying across the border and the Scottish–English border is hardly noticeable except for the big 'Welcome to Scotland' signs on one side of the road and the big 'Go Back to Scotland' signs on the other.

The reason I found myself in Port Jervis was that I was in my gap year. These days young people on their gap years tend towards Cambodia and Chile and Zambia, but in my day it was Port Jervis. They called it the Port Jervis trail where young people from all over the UK rejected the values of their parents and the Establishment and made a pilgrimage to an inland port on the old Delaware–Hudson Canal in upstate New York to liberate themselves socially, politically and sexually. To the generation that came of age in 1982 Port Jervis has all the resonance of Woodstock, Kathmandu, Ibiza, Glastonbury and Marrakech rolled into one.

Sadly, that is total pants. If I'm honest, I didn't really want to go to Port Jervis. In my defence it wasn't exactly my choice. I had decided that the cheapest way to get to the land of bacon and menthol was to volunteer to be a camp counsellor somewhere in America. Summer camps in the states have no equivalent in this country. They are mini-Guantanamos for children to be incarcerated in over the summer vacation. Americans who look upon our boarding schools with horror think nothing of sending their own children away for eight weeks to camps which brazenly advertise low standards of accommodation as 'outdoor lifestyle', gross health and safety violations as 'adventure' and systematic brutality as 'structured activity'.

There are many different styles of camps in the US, some of the five-star variety with unrivalled leisure facilities. It was this style of summer camp that I had in mind when I applied for the visiting counsellor programme. On the application form there were interminable questions about sporting skills with a long list of sports, from archery to yachting, and boxes to tick depending on whether you were competent, experienced or professional. They didn't have a box for 'embarrassing' so I skipped this entire section.

Looking back on it I think I may have written somewhere on the application form that I could run fast. It should therefore have come as no surprise to me that the only camp I was offered was Camp McDougal, a camp for 'socially disadvantaged' kids from the Bronx. I arrived at the camp and was shown to my hut, which was called Algonquin after the local but largely extinct Native American tribe. It seemed wherever I went in the States the local segmentation was based on Native American tribes possibly to remind people of where those tribes were before they were killed off or forcibly moved somewhere else.* I was immediately familiar with the layout of the hut as it was a mirror image of the huts in Stalag Luft III, from whence *The Great Escape* was made. I could have told you twelve different ways of escaping from that hut but fortunately the door was generally left open.

After a couple of days' training in self-defence, riot control and the safe disposal of needles, we were introduced to the kids. On reflection, I think we did better getting the 'socially disadvantaged' kids as I heard later from other counsellors that the 'socially advantaged' kids were unbearable. What the kids, who were mostly from inner-city projects, most wanted to do was to run around and make a lot of noise. I had specialised in exactly that for about fifteen years previously so we all got on like a house on fire. At the end of the

---

* My class at Elms Road Primary School was divided into Angles, Saxons, Jutes and Danes. Something tells me that's probably been changed since I was there.

day, when we retired to our huts, things got even better. I asked them what they wanted to do in the absence of a TV and after various facetious and impractical suggestions, they decided they wanted to hear stories, and most of all war stories.

That was a great moment for me. Had they asked for romantic fiction they would have been bitterly disappointed but with war stories we could have been sentenced to fifteen years in our hut and they would have had a new story every night. As the Falklands War had just finished and had lightly impinged on their little consciousnesses, I allowed myself the slight fiction that I was a decorated Harrier pilot/naval commander/SAS trooper freshly returned from that conflict. As you can imagine this went down very well with the boys, who couldn't believe their luck having an all-action war hero posted to their hut. There was one rather precocious child who asked some pretty tricky questions about some impossible military logistics I'd created for myself, but he soon found himself outside doing a lot of 'sentry duty'.

The days at camp passed peacefully. Being a poor camp it had almost no leisure facilities other than trees. This was standard holiday environment for me and I immediately instituted the famous Browning Long Marches for the kids. It's amazing how the lippiness of even the toughest kids faded after the first ten miles. It was then I began to understand my parents' passion for the Long March. Back in camp they introduced me to break-dancing, which was happening on the streets of the Bronx at the time, and in return I introduced them to cricket. It was a close call as to who was more confused and bewildered. Also, I don't actually know how to play cricket, which may have added to the confusion.

Being with kids all day, as any parent will tell you, is pretty exhausting. The counsellors did get some evenings off, though, and on these we liked to go down to a local bar on the nearest proper road. It was called My Buddy's Place and it was just how a bar should be. There was the big horseshoe bar in the middle serving cold beer and pizza. A few chairs, a television showing impenetrable American football and a large pool table. I had already started my under-age drinking in the UK insisting, when challenged, that I was twenty-five even before my voice had broken. In My Buddy's Place you had to be twenty-one to drink so I had an unusual opportunity to start my under-age drinking all over again.

In many states in America you're not allowed to have a drink until you're twenty-one. On the other hand, you are allowed to operate a surface-to-air missile and a range of other sophisticated weaponry shortly after your fifth birthday. The standard defence for this rather curious state of affairs is that 'guns don't kill you, people do'. The same logic doesn't seem to apply to drinking. 'Alcohol doesn't make you drunk, drinking it does.' In my view it's difficult to do either with neither. I suppose we shouldn't complain because at least the kids who do the mass shootings at high schools are generally sober.

My Buddy's Place was where I had my first game of pool. One of the American counsellors was a big white guy from the Bronx called Michael Goggins, who was more Irish than any Irishman I've ever met. I very rarely thought about the IRA when I was at home but Michael Goggins was an absolutely passionate supporter of the Republican cause and regarded me as being directly responsible for the potato famine in Ireland and an agent of British imperialism. Whatever the rights and wrongs of the Republican cause, he was a particularly unpleasant advocate of it and he did his best to make my life a misery. When we arrived in My Buddy's Place one day, he challenged me to a game of pool. I don't think I'd ever played before, probably because I could see that not only did it involve a ball but a stick also seemed to be important. I knew from football that the prospects of me personally connecting with a ball were slim, so getting a third-party stick to connect with one seemed totally impossible.

Michael Goggins insisted we play. I picked up the mallet/bat/cue and after laughingly asking him to remind me of the rules, we began to play. I can't remember whether he was any good or not. All I remember is that I was absolutely brilliant. I potted shots beyond the realm of physical and trigonometrical possibility. Every ball I touched shot straight into the hole in some kind of ecstatic home-coming. We played ten games and I won every single one of them with total, effortless abandon. I was wise enough to make sure I didn't make any kind of political or historical capital from this divine intervention on my side. Curiously, Michael Goggins left the camp shortly afterwards for 'compassionate' reasons. I always wonder whether the game of pool had anything to do with this. Since then I have played about another ten games of pool over twenty years or so and have lost every single one, and in many instances unzipped the baize with my cue and caused balls to leave the table and on some occasions the room. I don't know what happened in My Buddy's Place but it never happened again.

Most people who are interested in maps are also interested in the heavens above. Looking up at night is like looking at a giant dot-to-dot puzzle of the universe. Of course the moon and all the other celestial bodies have a powerful effect here on earth and this consti-tutes the study of astrology. We're the only advanced civilisation where educated people don't believe in this effect, but when you've had an intense karmic relationship with the astrologer of the *Daily Express*, as I have, you'll know what a powerful force it is. Adolf Hitler also had a complete dependence on his astrologer and wouldn't pass wind let alone invade Russia without checking that the stars were in propitious alignment. His astrologer very cannily told Adolf that Germany would emerge stronger and more powerful after a short and unpleasant learning phase but omitted to tell him that he wouldn't survive the lesson.

Love is famously star-crossed and I had a superb illustration of this when I was in Port Jervis. The camp was outside the town in some very lovely woodlands, studded with small lakes. I had been particularly taken with a lovely Dutch girl called Elisabeth de Wilde who was also a camp counsellor at Camp McDougal, despite her being exceptionally good at sport. On the evening of my nineteenth birthday I asked her to meet me after dark at the lake shore, which was strictly forbidden by all the higher authorities. Amazingly, she agreed. I then liberated a small rowing boat and rowed her out to the middle of the lake. My thinking was that, firstly, it was very romantic and, secondly, it would be difficult for her to escape. I had just started the whispering romantic phase as we drifted in the middle of the lake when I heard a splash nearby. This was followed by a number of other small splashes and some larger ones. I had no idea what this was and I begin to think that the entire camp, alerted to our plans through some basic slip in security, had ringed the lake and were now throwing stones at the boat in order to distract or possibly sink us.

Elisabeth told me the splashes were probably fish jumping for insects. This seemed like a reasonable explanation and I only wished I hadn't shared my own ludicrous one with her. Fortunately my embarrassment was masked by a major cosmic event. Huge pillars of light seemed to be probing the skies ahead of us. I immediately assumed this was part of the camp's organised search party that would undoubtedly be looking for curfew breachers like us but Elisabeth told me that they were the Northern Lights. Whatever they were, they were pretty impressive and an excellent aid to rekindling a romantic atmosphere. As it turned out this was just a celestial warm-up for what was to come. We rowed back and lay on the shore looking up at the night sky. I tried a few chat-up lines and the universe, suitably impressed, immediately provided a meteor shower of incredible length, brightness and intensity. On each shooting star we made a wish and pretty much all of them came true save one of my latter ones which had overambitious financial targets attached.

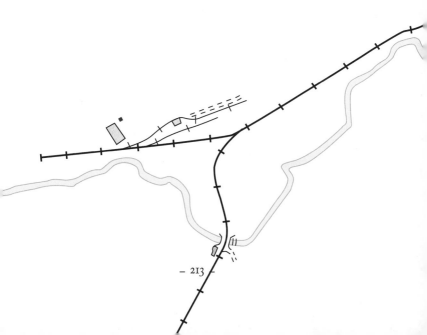

BAR CHART SHOWING HIGHS AND LOWS
OF MY TIME IN NEW YORK CITY

1. Incredible rudeness from U.S. immigration officials
2. Welcomed to Miguel's apartment
3. Accidentally swore at Miguel's mother
4. Visit to Fergus in South Bronx
5. Mugged in Lower West Side
6. Disco heaven at Studio 54 nightclub
7. Learning moonwalk with Michelle
8. First public performance of moonwalk
9. Breakfast at comfort diner with crispy bacon
10. Purchases at The Wiz record shop, the Bronx
11. Extraordinary night of passion with Elisabeth de Wilde
12 Feedback from Liz about night of passion
13. *Papillon* with Steve McQueen & taxi ride in the rain
    with Elisabeth de Wilde
14. Flight home

CHAPTER 21

New York City, New York State, USA

If you haven't visited New York before, don't make the mistake of thinking that it's going to be in any way like the old York in York-shire. They have pretty much nothing in common. Before it was New York it was called New Amsterdam by the Dutch and bears even less resemblance to that great city. When you've sailed halfway around the world to found a new city in a new continent, why not spend a little time thinking of a new name? Mind you, calling New York New York was better than just calling it York, like the old one. Whenever I try to fly from Birmingham to Edinburgh the travel site asks me whether I would prefer to depart from Birmingham, England, or Birmingham, Alabama. I sometimes suspect they've programmed sarcasm into that particular website.

Nevertheless, don't let the fact that New York doesn't have a Viking Centre put you off. It has many other remarkable features, the most impressive of which is its sheer verticality. Most New Yorkers when they go home, go up. Elevator etiquette is an art form in New York and natives can adjust their small talk to the exact time required to get to their interlocutor's floor. Pretty much all the friends I made at Camp McDougal lived in the Bronx and most lived in tower blocks, or projects as they call them.

My best friend was a Puerto Rican called Miguel Lopez. He was the most dapper man I have ever met. Some people in life choose to play the double bass and are forever destined to carry around a huge case with them. Miguel had chosen never to wear anything without razor-sharp creases and was forever destined to travel with an ironing board. All through the summer at Camp McDougal Miguel never made any concession to the fact that we were in a remote rural location with very little in the way of tarmac. He wore immaculately ironed street gear even when we were supposed to be doing a nature walk through the thickest and dirtiest part of the countryside. I remember him exercising all sorts of contortions and rapid movements to keep his trousers clear of thorn bushes and bogs. I think this may have been the origin of break-dancing.

Miguel invited me to stay with him in the Bronx when summer camp was over. He lived in a tiny apartment with his mother, who was a first-generation immigrant. Her English was at about the same level as my Spanish. For my part, thanks to my time in Texas and my recent refresher course with some of the Hispanic kids at camp, I could now swear fluently in Spanish. As the rest of my Spanish was very shaky this made conversations difficult, if not dangerous. I believe one of the highlights was when I thanked her for a meal that was 'an absolute bastard'.

One day Miguel took me to see one of his best friends whose name, unless I'm very much mistaken, was Fergus. He lived in another project block half a mile away and, as we approached the building, a big lady hung over her balcony on the second floor and shouted down, 'Yo, Miguel, who's that albino with you?' He explained I was from England, which seemed to be enough to explain any kind of aberration in behaviour or appearance on my part. We went up in the lift/elevator to see Fergus. When the doors opened on Fergus's floor I immediately heard a colossal thumping sound as if a wrecking ball was trying to knock this particular floor out of the building. It turned out that the noise was coming from Fergus's apartment. We opened the door and the noise, if it was possible, got louder. Fergus himself was sitting on a sofa between two speakers each of which was bigger than the sofa itself. On the seat next to him were a number of beer bottles and a very full ashtray. In the apartment was an over-whelming smell of something that wasn't necessarily tobacco.

I can't remember what Fergus was listening to but thankfully it had intermittent moments of silence between the big beats. We managed to get the introductions done in those micro-seconds of quiet and, as if in pre-emptive apology, it was explained that I was English. In the next break between tracks Fergus leant over and asked:

'Do you know Prince Charles?'

I didn't have time to say 'no' before the next track started so I had five minutes to think of the answer which, when it eventually came, was:

'Yes, he's my homeboy.'[*]

'Cool,' was Fergus's reply. Then he put on a twelve-inch party mix of 'Infatuation' by Up Front. As this meant conversation could not now resume for at least ten minutes we made our excuses, did some rather complex handshaking and made our exit.

My other close friend was Tyrone Randal, who sounded as if he was from County Donegal but was in fact a six-foot-seven black man also from the Bronx. Tyrone and I played a lot of basketball together. I was his ideal partner. I couldn't catch the ball or throw it but I could run around tirelessly for hours trying to get in his way while he executed a range of dazzling airborne manoeuvres which all resulted in the ball whistling through the net. What we also had in common was our taste in music. At Camp McDougal I was very glad that I had taken the lonely and less trodden path that leads to disco because we had something to talk about. Deep Purple had never made much impression on me or the South Bronx. That was

---

[*] Best friend.

the summer of the *Thriller* album from Michael Jackson. The neighbourhood was alive with it and I remember dancing to it in the streets with the sound of 'Beat It' in one ear and the sound of most of the population of the South Bronx laughing themselves sick at my English dancing in the other.

One of the other counsellors I met at Camp McDougal was a tall black girl called Michelle who took pity on my dancing. She invited me round to her house, which turned out to be about thirty floors up somewhere in Manhattan. Her parents ran a bakery and were both as white as I am. I'm pretty sure Michelle was adopted but I couldn't get over the impression that her parents were really black but were just permanently covered in flour. Michelle taught me how to dance to Michael Jackson. Back then I was grateful but now, when I display this unusual talent, it seems to get me fewer cool points than you'd think. I also pull various hamstrings, vertebrae and hernias and have to cut short the full routine before I am hospitalised.

It was on the streets of the South Bronx that I became acutely aware of the presence of drugs. Fortunately by this stage of my life there was very little I didn't already know about them. Before we set out on our Texan adventure, when I was about twelve years old, my father was very concerned that we should be aware of the drug menace that we would inevitably encounter over there. This concern arose, not from his knowledge of junior high schools in Texas, but from his expert knowledge of agriculture in South America. He'd seen the crops being grown there in a volume that far exceeded local consumption and wanted to avoid the final destination being the nostrils of the young Brownings.

To this end my father approached our drug education in a similarly rigorous and methodical way as he had our sex education. In the absence of any helpful books entitled *Everything You've Always Wanted to Know About Drugs But Were Too Out of It to Ask*, he had to come up with his own. Being an academic he did some in-depth research, which I'm assuming was done in libraries rather than in crack dens in Botley. The presentation we received on drugs and the shocking menace thereof was comprehensive, to say the least. In a lecture that must have lasted the best part of two hours, my father listed every drug known to man, their normal method of application (smoking, sniffing, injection, suppository etc), the very dubious short-term benefits and the absolutely appalling medium-to-long-term consequences of usage.

I'm pretty sure my father never really grasped the difference between recreational and therapeutic drugs because, as far as I'm aware, not much in the way of recreational drugs gets delivered by suppository, topical cream or IV drip. We must have covered a range of chemotherapy, antibiotics and indigestion remedies in among the more commonly recognised hallucinogens. In the end, this actually proved a remarkably efficient way of putting us off drugs as it left us all with the impression that a quick toke on a spliff (my father went to great lengths to use the correct druggie slang current on the streets and in GPs' surgeries) would be the high road to cancer and irritable bowel syndrome, if not both.

As it turned out, I never got so much as a sniff or even a whiff of drugs in Texas. The only illicit substance seemed to be chewing tobacco which Jim Hickman and the mastodons used to place under their bottom lip. This simultaneously gave them a nicotine rush and also made them look even more prehistoric than they already did. In the South Bronx, however, we were all slightly older and the story was a little bit different. The group I was hanging out with talked about 'candy' a lot, which to begin with I thought referred to sweets. I have a sweet tooth and I may have surprised them with my initial enthusiasm to get hold of some candy. Once I understood what they were actually talking about I had to do some Tour de France-strength backpedalling.

My friends took me to some interesting places to savour the local pharmaceutical delicacies and each time I politely demurred, citing my close relationship with Prince Charles and the fact that, were I to be caught taking drugs, it would precipitate a constitutional crisis in my country. This was no more eccentric than most of the other stuff I was telling them, so I managed to survive most of the sessions on nothing harder than a Peppermint Pattie which was indeed candy, a circular after-dinner mint as big as a Wagon Wheel.

One evening we went to a shop that seemed to be completely boarded up. A door in the middle of the shop let into a small, caged area, like the foyer of a particularly paranoid sort of bank. A small door was opened and the 'candy' was passed out. I thought this was all terribly exciting, and that I was living out an episode of *Starsky and Hutch*, when a little girl of no more than nine walked in behind us, gave her order and then marched out with her assorted pharmaceuticals as if they were nothing more than a Sherbet Dip Dab.

I thought about this little girl exactly twenty years later when I was running the New York Marathon. The marathon is famous for going through all of New York's five boroughs and, if you've got four or five hours to spare, is actually a very good way of seeing the city, especially bits you wouldn't normally see on the tourist trail. It's an amazing event, the New York Marathon, and one of the most amazing things to me was that before the start they had huge tables piled high with donuts from which any runner could help themselves. I hadn't seen so many donuts in one place since my time of prayer and reflection at the Church of the Good Shepherd in Austin, but I resisted temptation because it's actually quite hard to run twenty yards after a few donuts let alone twenty-six miles.

The marathon went pretty well for me and I was running strongly until I got to about twenty miles, when I hit the infamous wall. My body had run out of sugar, it didn't want to go any further, and it made that clear to me in no uncertain terms: that is to say, it completely stopped working. It just so happens that twenty miles into the New York Marathon is the South Bronx. Spectators are few and far between on that part of the course but there was a little girl and her mother handing out candy to runners. Before I took them, a little warning voice cautioned me about little girls with candy in the Bronx and that if I wasn't careful I might be rocket-powered for the rest of the race and fail every drug test known to man. But I was too tired to care so I took a handful of candy from her, thanked her kindly and, after quickly crunching my way through them all, managed to run the rest of the way to the finish in Central Park.

A mile before the finishing line, I ran past a place in the park resonant of another memory, equally sweet. While I was staying in the Bronx, I had met up again with Elisabeth de Wilde, the beautiful Dutch girl from Camp McDougal. We wandered around aimlessly as poor young couples do and, as the forecast was not good, we decided to go to the cinema. I was young and foolish and thought you should go to the cinema to see a film you really wanted to see rather than just pay for any old film so you could snog throughout. We saw *Papillon*, a fantastic escape movie (there's nothing like an escape movie for sheer escapism). I think I lost some brownie points in the cinema because I was absolutely transfixed by the film and was probably concentrating rather too exclusively on the screen. In my defence you also have to remember that the hero of this story is Steve McQueen with whom I share a remarkable similarity in the way we look, the way we carry ourselves and our incredibly cool way of dealing with attractive women and uncooperative men. No one else has made this connection but then both Steve and I are pretty enigmatic figures and it takes time for the uncanny resemblance to become apparent to most people, often a very great deal of time.

On the way out of the cinema the sky was black. As it was still the afternoon, this was not a good sign. We decided to get a cab back across town. Luckily it was one of the old Checker cabs with a cavernous passenger compartment and together we snuggled down in the back. We'd just entered Central Park when the heavens opened. When I think about my time with Elisabeth I would have to describe it as phenomenal simply because we couldn't so much as hold hands without nature putting on some spectacular show by way of a romantic background.*

In this particular case, the storm was ferocious in its intensity. The din of the rain hammering against the roof of the taxi was incredible and pretty soon we could go no further in the park because the road ahead was flooded where it dipped under a bridge. So we stayed in the back amidst the lightning and thunder and drumming rain and I attempted to retrieve some of the brownie points lost in the cinema. If there were blue plaques commemorating the happiest moments in people's lives, then there would definitely be one in the middle of Central Park, on the side of the road, just before the bridge. My name would be on it and, I hope, Elisabeth's too.

One of the nice things about that summer in New York – and there were a lot of nice things – was that the Fatted Calf had also spent the summer in America and we arranged to fly back together. When we met he had the look of someone who'd just made love very successfully to someone very beautiful. His entire summer had been spent making love to America, which was everything he'd dreamed a country could be: an efficient market system of supply

---

* I often wonder what happened to Elisabeth when she returned to the Netherlands and whether her every move is still accompanied by severe weather warnings. I would love to have seen what nature came up with when she really fell in love. But then some other lucky man will have had that pleasure.

and demand, abundant natural resources, a light regulatory environment and unlimited crispy bacon. He told me that one day he'd return and make America his own.[*]

We spent a couple of days in Manhattan together. We bought a few records and some clothes (the Fatted Calf gave me what he called a 'sub-prime' loan) but mostly the Fatted Calf seemed content to wander round the financial district in the bright sunshine, soaking up the petrol-like smell of unfettered capitalism. We went to the Twin Towers of the World Trade Center and I remember him standing at the very root of one of the towers and hugging it. I'm glad he did.

Finally, at the end of a long, long summer, it was time to fly home to England. It's a shame that the US Immigration officials who 'welcome' you on your way in aren't there to say goodbye on your way out because then the world could tell them exactly what they think of their appalling manners. Sadly, they were too busy abusing people in the Arrivals hall so the Fatted Calf and I just stocked up on menthol shaving cream and headed for home.

[*] He did, and was responsible for the huge growth in the American economy in the 1990s and early 2000s.

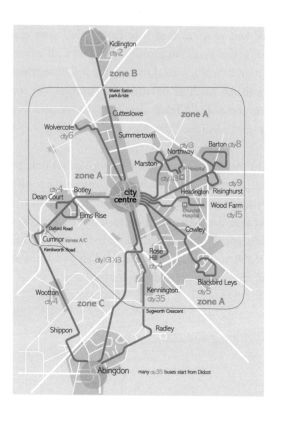

Kidlington
city2

zone B

Water Eaton
park&ride

Cutteslowe

zone A

Wolvercote
city6

Summertown

city13
Northway

Barton city8

Marston

Headington

Risinghurst city9

Headington

Wood Farm
city15

zone A

Botley

Dean Court city4

city centre

Elms Rise

Churchill
Hospital

Cowley

Oxford Road

Cumnor zones A/C

Kenilworth Road

cityX3 X13

Rose
Hill
city4

Blackbird Leys
city5

zone A

Wootton
city4

zone C

Kennington
city35

Sugworth Crescent

Shippon

Radley

Abingdon   many city35 buses start from Didcot

Epilogue

School and childhood were now over for me, although many years of immaturity lay ahead. The Fatted Calf was already at the London School of Economics doing some ground-breaking work on the exploitation of child labour (I think there's a picture of me in his dissertation). I knew it was now time for me to leave home, to stretch my wings, find my feet and head for new, uncharted territories. The autumn arrived and finally I took the first step on my next great adventure. Off I went to the world-famous University of Oxford. There's a bus from Botley.

Acknowledgements

p. 60 Jamaica Farewell. Words and music by Irving Burgie. Copyright © 1955; Renewed 1983 Cherry Lane Music Publishing Company, Inc. (ASCAP), Lord Burgess Music Publishing Company (ASCAP) and Dimensional Music of 1091 (ASCAP). World-wide rights for Lord Burgess Music Publishing Company and Dimensional Music of 1091 administered by Cherry Lane Music Publishing Company, Inc. International Copyright Secured. All rights reserved.

p. v Weather map from the *Daily Telegraph* November 30, 2007 by courtesy of AccuWeather, Inc. 385 Science Park Road, State College, PA 16803 (814) 237-0309 © 2008

p. 1 Chipping Norton © Bish

pp. 11–12 Niagara Falls and Neighbouring Countries reproduced by permission of University of Texas Libraries, Austin, Texas

p. 24 El Salvador © Rand McNally–R.L.08–S–60

p. 36 Kidlington, Oxfordshire © Bish

p. 46 Kidwelly, South Wales reproduced by permission of Ordnance Survey on behalf of HMSO. © Crown copyright 2008. All rights reserved. Ordnance Survey licence number 100012246

p. 56 Jamaica © Publisher

p. 66 *Oxford From North Hinksey* by Joseph Mallord William Turner © Manchester City Art Gallery/Bridgeman Art Library

p. 78 Loughton, Essex reproduced by permission of Ordnance Survey on behalf of HMSO. © Crown copyright 2008. All rights reserved. Ordnance Survey licence number 100012246

p. 90 Harwell Hill © Publisher

p. 102 Heidelberg Castle from *A Guide to Heidelberg: Its Castle & Environs* published by Johannes Hörning, Heidelberg

p. 112 Mother's Mental Map of the Isle of White © Bish

p. 122 Distribution of Native Tribes in Texas © Cherokee Publications

p. 132 Six Flags over Texas and Six Favourite Flags

p. 142 White Sands National Monument reproduced by permission of National Park Service, U.S. Department of the Interior

p. 154 Confident Young Man from Botley © Bish

p. 164 Andalo, Italy ski map from SciClub Carpenedolo

p. 174 Allied invasion plan for Normandy, France

p. 184 Geological map of Spain © Instituto Geológico y Minero de España and Sociedad Geológico de España

p. 194 Dangerous stretches of the River Thames around Oxford © Bish

p. 204 Poer Jervis, New York meteor shower, 19th century engraving

p. 214 Bar chart of highs and lows in New York City © Publisher

p. 224 Oxford Bus Company route map © Best Impressions and Oxford Bus Company